Time-Series

Other statistical books by Sir Maurice Kendall

Multivariate analysis
Rank correlation methods
Exercises in theoretical statistics
A course in the geometry of *n* dimensions*

With Professor A. Stuart
The advanced theory of statistics (3 volumes)

With G. Udny Yule
An introduction to the theory of statistics

With Professor P. A. P. Moran
Geometrical probability*

Works edited by Sir Maurice Kendall
Mathematical model building in economics and industry (2 volumes)

With Professor A. Stuart
Statistical papers of George Udny Yule

With Professor E. S. Pearson
Studies in the history of statistics and probability

* A volume in "Griffin's Statistical Monographs and Courses"
  Full descriptive list available

Time-Series

Sir Maurice Kendall

M.A., Sc.D., F.B.A.

Second edition

Quis separabit nos

Charles Griffin and Company Ltd

London and High Wycombe

CHARLES GRIFFIN & COMPANY LIMITED
Registered Office:
5A Crendon Street, High Wycombe, Bucks HP13 6LE

First published   1973
Second edition   1976

ISBN: 0 85264 241 5

Set by E W C Wilkins Ltd, London & Northampton
Printed in Great Britain by Butler & Tanner Ltd, Frome

Preface to the First Edition

In the last thirty years the theory of time-series has been transformed into a new subject. In part this is due to the introduction of probabilistic ideas into what was formerly treated deterministically; in part it is attributable to the power of the electronic computer, which has removed the obstacles imposed by the extensive and tedious calculations involved in most time-series studies. There has, nevertheless, tended to appear a rift between sophisticated theory and practical application, and although there exists an extensive literature in scientific and professional journals there are few books which attempt to treat the subject in its entirety for the benefit of the practising statistician.

That is my reason for writing this book. It aims to present the basic ideas and techniques of the subject, with as much exemplification as space will permit and a determination not to let the mathematics multiply beyond necessity. I have tried to make it the sort of book that I would like to have had put in my hand when I first became interested in time-series many years ago.

I am indebted to Professor Dudley J. Cowden and the Director of the School of Business Administration, University of North Carolina, for permission to reproduce Appendix Table A; to Professor James Durbin and the Editors of *Biometrika* for permission to reproduce Appendix Tables B; to Professor C.W.J. Granger for permission to reproduce Fig. 8.4; and to Dr D.J. Reid for permission to reproduce Fig. 9.1.

M.G.K.

London
February, 1973

Preface to the Second Edition

A number of misprints and obscurities have been removed and some references added in this edition, which otherwise follows the same lines as the first.

M.G.K.

London
December, 1975

Contents

1

General ideas

1.1 Time is perhaps the most mysterious thing in a mysterious universe. But its esoteric nature, fortunately, will not concern us in this book. For us, time is what it was to Isaac Newton, a smoothly flowing stream bearing the phenomenal world along at a uniform pace. We can delimit points of time with ease and measure the intervals between them with great accuracy. In fact, although errors and random perturbations of most of the variables with which we shall be concerned are frequent and important, we shall only exceptionally have to consider errors of measurement in time itself. Such difficulties as arise in measuring time-intervals are mostly man-made (e.g. in the absurd way in which the Christian world fixes Christmas Day but allows Easter Sunday to vary over wide limits).

1.2 From the earliest times man has measured the passage of time with candles or clepsydras or clocks, has constructed calendars, sometimes with remarkable accuracy, and has recorded the progress of his race in the form of annals. The study of time-series as a science in itself, however, is of quite recent emergence. The recording of events according to a horizontal axis in which equal intervals correspond to equal spaces of time must have occurred a thousand years ago; for example, the early monkish chants recorded on the eleven-line musical stave are a form of time-series. In this connection Fig. 1.1 (from Funkhauser, 1936) may be of some interest as the earliest diagram known in the Western world in which the essential concepts of a time-graph are apparent. It dates from the tenth, possibly eleventh century, and forms part of a manuscript consisting of a commentary of Macrobius on Cicero's *In Somnium Scipionis*. The graph was apparently meant to represent a plot of the inclinations of the planetary orbits as a function of the time. The zone of the zodiac is given on a plane with the horizontal (time) axis divided into thirty parts and the ordinate representing the width of the zodiacal belt. The precise astronomical significance need not detain us; the point is that even at this early stage the time-abscissa and the variable-ordinate were in use, albeit in a crude and limited way.

1

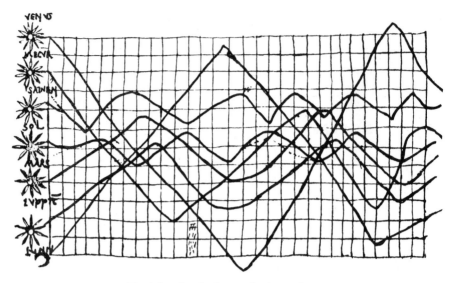

Fig. 1.1   Graph of an early time-series

1.3   Notwithstanding the invention of coordinate geometry by Descartes,
the pictorial representation of time-series was a late development. As recently
as 1879, Stanley Jevons, whose book on *The Principles of Science* was by no
means intended for schoolboys, felt it necessary to devote some space to the
use of graph paper. Possibly the first (and certainly one of the earliest) writers
to display time-charts in the modern way was William Playfair, one of his
diagrams being reproduced in Fig. 1.2. (The diagram was published in 1821.)
Playfair, the brother of the mathematician known to geometers as the author
of Playfair's axiom on parallel lines, makes a number of claims to priority in
diagrammatic presentation which, whether justified or not, at least demonstrated that the procedure was unfamiliar.

1.4   In the nineteenth century theoretical statistics was not the unified subject it has since become. Work went on in the physical sciences very largely
independently of work in economics or sociology, and at that period the
ideas of physics were entirely deterministic; that is to say, a phenomenon
tracked through time was imagined as behaving completely under deterministic
laws. Any imperfections, any failure of theory to correspond with fact was
either dealt with by modifying the theory in a deterministic direction (as,
for example, in the discovery of Neptune) or attributed to errors of observation. When, during the latter part of the nineteenth century, attempts were
made to apply in the behavioural and biological sciences the methods which
had been so successful in the physical sciences, this deterministic approach
was taken over, together with the rest of the mathematical apparatus which

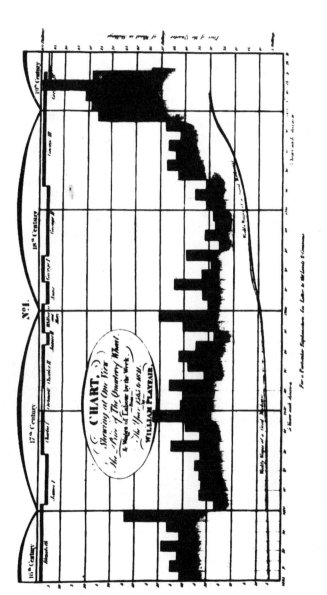

Fig. 1.2 Chart combining a graph and a histogram : from Playfair's *A Letter on our Agricultural Distress* (1821)

had been developed in the latter. At that point the modern theory of statistics began with the realization that, although individuals might not behave deterministically, aggregates of individuals were themselves subject to laws which could often be summarized in fairly simple mathematical terms.

1.5   Time-series, however, resisted this change of outlook longer than any other branch of statistics. Up to 1925 or thereabouts, a time-series was regarded as generated deterministically, and the evident departures from trends, cyclicality or other systematic patterns of behaviour which were observed in Nature were regarded as "errors" analogous to errors of observation. They occurred all too frequently in some fields, but were regarded in much the same way as an electronics engineer regards "noise", as a fortuitous series of disturbances on a systematic pattern. In particular, fluctuating phenomena such as the trade "cycles" of the nineteenth century were subjected to Fourier analysis as if they were the sum of a number of harmonic terms generated by oscillators of the classical kind. The failure of such models to account for the features of much fluctuation observed in such things as trade, population and epidemics, though disappointing, did not put a stop to the search for underlying cyclical movements of a strictly harmonic kind; nor, in fact, is the belief in their existence yet dead.

1.6   It was in 1927 that Udny Yule (see his *Statistical Papers*, 1971) broke new ground by a fertile idea which forms the starting point of much of the study of time-series analysis which has been carried out since that time. Working on sunspot numbers, which obviously fluctuate in a manner that cannot be entirely due to chance, Yule was struck by the fact that the amplitudes of his series and the distances between successive peaks and troughs were irregular. The illustration which he used to explain his fresh approach is classical: if we have a rigid pendulum swinging under gravity through a small arc, its motion is well known to be harmonic, that is to say, it can be represented by a sine or cosine wave, and the amplitudes are constant, as are the periods of the swing. But if a small boy now pelts the pendulum irregularly with peas the motion is disturbed. The pendulum will still swing, but with irregular amplitudes and irregular intervals. The peas, in fact, instead of leading to behaviour in which any difference between theory and observation is attributable to an evanescent error, provide a series of shocks which *are incorporated into the future motion of the system*. This concept leads us to the theory of stochastic processes, of which the theory of stochastic time-series is an important part, but only a part. Its usefulness will be exemplified many times in the sequel.

1.7   The characteristic feature of time-series, in contradistinction to other statistical subjects, is that the observations occur in temporal order, which is not quite such a trite remark as it sounds. The implication is that we shall, among other things, be interested in the relationship of values from one term to the next, in the serial correlations along the series. When we have several series to consider as a multivariable complex, there will thus arise a dimension of complexity beyond that of multivariate analysis. In the latter, as ordinarily understood, we are concerned with relationships or inter-relationships among variables, regardless of the order in which the individuals which bear them are presented to us. With a multivariable time-series we have to consider, in addition, the correlations and cross-correlations among the series when one or more lead or lag behind the others.

1.8 In passing, we may remark that there are still further dimensions of complexity awaiting statistical study. The techniques we apply to time-series, essentially unidimensional in time, can be extended to spatial situations wherein more than one time-like dimension, so to speak, is involved. For example, we may be interested in the intensity of wire-worm infestation in a field − a two-dimensional situation; or in the variation of wind velocity in the upper atmosphere − a three-dimensional situation; but general theories in such cases become very complicated and tend to lose touch with reality. We shall be sufficiently exercised to handle the one-dimensional movement through time.

Aggregation

1.9 Some variables exist at every point of time and can be regarded as having a continuous existence; such, for example, are the temperature at a given place, the price of a given commodity on an open market, or the position of a projectile. Others exist only in virtue of aggregation over a period; for example, rainfall, industrial production, miles travelled by an airline. Some, again, exist (or come into existence) only at discrete time-intervals: the yields of a given crop at harvest, the majority of a political party at a general election, rents payable on quarter days. Sometimes we have no choice in the time-points when observations are carried out, as in harvest yields. In others we have some choice, but not an unlimited one; for instance, imports are usually provided by month, and only exceptionally would a Government agency be willing to give figures for shorter intervals or split periods. In extreme cases we can observe the series almost continuously, or at any rate, as long as our patience and effort will last; records on a rotating drum of temperature or barometric pressure, or the alpha rhythm of the brain on an encephalograph are familiar examples.

1.10 We shall occasionally have to give some attention to the choice of the time-points of observation in cases where we have any choice. There is often some conflict of interest in selecting the best set of points. For reasons of economy we usually do not want to multiply the observations beyond necessity; on the other hand, we do not want them to be so sparse that we miss some essential features of the situation under study. If we are interested in seasonal fluctuation we must take observations at several points in the year, whereas if we wish to ignore seasonal variation it may be better to take observations at one date in the year or to aggregate (when this is permissible) over the whole year. In time-series analysis, as we shall see on more occasions than one, there are rarely rules of general and universal application; a great deal depends on the purpose of the study.

Example 1.1 (Working, 1960)

As an exámple of the effects that can arise from aggregated series, consider the case when the increments in the series from one time-period to another are values of a random variable ϵ with zero mean. It has been suggested that some price movements, for example on the Stock Exchange, are of this character, and certainly they bear a close resemblance to a so-called "wandering series". Thus the first differences of the series are random.

Suppose, then, that the series of the first m terms is

$$u_1, u_1 + \epsilon_1, u_1 + \epsilon_1 + \epsilon_2, \ldots, u_1 + \epsilon_1 + \epsilon_2 + \ldots + \epsilon_{m-1}$$

and the second set of m is

$$u_1 + \epsilon_1 + \ldots + \epsilon_{m-1} + \epsilon_m, \ldots, u_1 + \epsilon_1 + \epsilon_2 + \ldots + \epsilon_{2m-1}. \quad (1.1)$$

If we average the first set and subtract from the average of the second, we find for the first difference of the averages (which is equivalent to m times the first differences of the cumulated series)

$$d = \frac{1}{m} \{ \epsilon_1 + 2\epsilon_2 + \ldots + (m-1)\epsilon_{m-1} + m\epsilon_m + (m-1)\epsilon_{m-1} + \ldots + \epsilon_{2m-1} \}. \quad (1.2)$$

The variance is

$$\text{var } d = \text{var } \epsilon \cdot \frac{1}{m^2} \left\{ \sum_{j=1}^{m-1} j^2 + \sum_{j=1}^{m} j^2 \right\} = \frac{\text{var } \epsilon}{3m}(2m^2 + 1).$$

The covariance of successive first differences, on multiplying (1.1) by itself advanced m units in t, is

$$\text{cov} = \text{var } \epsilon \cdot \frac{1}{m^2} \left\{ \sum_{j=1}^{m-1} j(j-1) + m(m-1) \right\}$$

$$= \frac{m^2 - 1}{6m} \text{ var } \epsilon. \quad (1.3)$$

Thus the correlation between successive differences of the cumulated series is

$$\frac{m^2 - 1}{2(2m^2 + 1)}. \quad (1.4)$$

Thus, although the first differences of the original series are independent, those of the cumulated series (or averages) are not. If $m = 12$, the correlation is about 0·25, so that even if differences from month to month were independent, the differences of the cumulated series from year to year would not be.

Continuity and discontinuity

1.11 In the sequel, when I speak without qualification of a continuous or a discontinuous time-series I am referring to the continuity of the variable, not the continuity in time. Thus the production of aircraft is a discontinuous variable; crop yields are a continuous variable, notwithstanding that they occur at discontinuous time-points. Strictly speaking, human population in a given country would, on this convention, be discontinuous, but the numbers are usually so large that the variable can be regarded as continuous without serious loss of accuracy.

1.12 There is another sense in which the word "continuity" is used in connection with time-series, especially those series which consist of index-numbers based on varying weights. An index-number of stock-exchange prices will illustrate the point. To construct such an index, say of the most important shares, we have to assign weights to the constituents on some such basis as the total capital of the companies concerned at some fixed base-point of time. These weights alter as time goes on and, when they become too obsolete, have to be updated. The question then arises whether the new series is "continuous" with the old. In practice a certain amount of "continuity" is achieved by splicing the index-numbers at the point of change. Indices of share prices, as a matter of fact, have to be updated in a small way constantly because certain shares drop out of (e.g. from takeovers or amalgamations), and others come into, the most important set, however defined. One therefore meets in the literature with expressions like a "continuous series" which do not describe continuity in time or measure, but only continuity in the sense of being constructed on a roughly comparable basis over the period concerned.

Calendar problems

1.13 For the most part we shall consider series which are observed at, or aggregated over, equal intervals of time. In physical, biological or meteorological studies we can usually ensure that observations shall be of such a kind. In economics, however, and indeed in behavioural subjects generally, the matter stands differently, and there are a number of nuisance effects which have to be attended to before we can apply sophisticated techniques to the series. For some of these effects we can blame Nature, e.g. in not arranging that the solar year shall consist of an integral number of days. Most of them, however, are man-made, particularly those relating to the calendar. The inequality in the length of months is one such case; the fact that a month may include either four or five week-ends is another. Movable feasts and public holidays contribute their own share of confusion, especially Easter, which may fall into either the first or the second quarter of the year. Even series derived from experimental observation on the factory floor may suffer from imperfection, both in the large (as, for example, due to strikes) or in the small (as, for example, due to meal-breaks).

1.14 The methods by which, in a useful colloquial phrase, we "clean up" the series vary according to circumstance and opportunity. We may briefly note a few of them:

(a) For figures of production by calendar month we can obtain a certain measure of comparability by correcting to a standard month of 30 days, e.g. by multiplying the production in February by 30/28 and that of March by 30/31. (It would be better to abstract the production of the 31st January from that month and add it to February, and likewise for the 1st March, but this is often administratively impracticable.) One must remember that the total of twelve such "corrected" months may not be exactly equal to the annual production, even if "corrected" to a year of 360 days.

(b) A similar comparability for industrial figures may be obtained by correcting for the number of working days in the month.

(c) Short-term effects may sometimes be eliminated by aggregation. If it is sufficient, for example, to work with six-month rather than with three-month periods, variations due to a movable Easter are no longer important. Average prices on a market which closes over week-ends may avoid gaps in information by averaging over a month; and so on.

(d) Data relating to values are particularly open to question because of changes in the value of money. There seems no better method of eliminating such changes than by dividing the data by some index measuring changes in money-values.

The length of a time-series

1.15 When we speak of the "length" of a series we are usually referring to the elapsed time between the recorded start and finish. But the word can also refer to the number of observations, so that, for example, a series of monthly rainfall figures over ten years would be described as of length 120. This usage would be avoided by the purist. The series would be of length 10, 40, 120 or about 3652, according to whether we observed it at intervals of one year, a quarter, a month, or a day.

1.16 A more important point concerns the amount of information in the series as measured by the number of terms. In ordinary statistical work we are accustomed to thinking of the amount of information in a random sample as proportional to the size of the sample. Whether this is a correct usage of the word "information" is an arguable question, but it is undoubtedly true that the variance of many of the estimates which we derive from random samples is inversely proportional to the sample size. This idea needs modification in time-series analysis because successive values are not independent. In colloquial terms, a series of $2n$ values (even if extending over twice the time) does not tell us twice as much about it as one series of n values. And if we sample a given length of series more intensively by observing it at half the interval of observation, and hence doubling the numbers of observations, we do not necessarily add very much to our knowledge if successive observations are highly positively correlated. The consequence is that n, the number of observations, is not, by itself, a full measure of the information. We shall see in the sequel that the precision of estimates which we can make from data involves, in general, the internal structure of the series in addition to n.

Some examples of time-series

1.17 The theoretical treatment to be developed later will need to be exemplified on some practical series which are now presented. Table 1.1 gives the yields of barley in England and Wales for the 56 years 1884-1939. Table 1.2 gives the sheep population of England and Wales for the 73 years 1867-1939. Table 1.3 gives the miles flown by British airlines for the 96 months January 1963–December 1970. Table 1.4 gives the immigration into the United States for the 143 years 1820 to 1962. Table 1.5 gives, for certain U.S. hospitals, the number of births of babies according to the hour at which they were born. Table 1.6 gives for 1960-1971 the quarterly average index of share prices on the London exchange as compiled by the *Financial Times*.

1.18 The series are presented in graphical form in the appropriate diagrams. They are fairly typical of the kind of material which we have to handle. Barley yields, by their very nature, occur once each year. The sheep population, though continuously existent, is observed only once a year at a fixed date (June 4th) so that seasonal movements are omitted. The airline data present a characteristic pattern of seasonal variation on a rising trend. Immigration, also on an annual basis, shows fluctuations, some of which can be identified with events such as war. The birth data are exceptional in that we have graphed the square root of the number rather than the number of births themselves (on the grounds that births probably follow a Poisson distribution and the square root transformation brings the variance nearer to a constant), and in that form they reveal a remarkable cyclical pattern. The F.T. index-numbers are typical of fluctuations in the stock market over a period of time.

Table 1.1 *Annual yields per acre of barley in England and Wales from 1884 to 1939 (data from the Agricultural Statistics)*

Year	Yield per acre (cwt)	Year	Yield per acre (cwt)	Year	Yield per acre (cwt)
1884	15·2	1903	15·1	1922	14·0
85	16·9	04	14·6	23	14·5
86	15·3	05	16·0	24	15·4
87	14·9	06	16·8	25	15·3
88	15·7	07	16·8	26	16·0
89	15·1	08	15·5	27	16·4
90	16·7	09	17·3	28	17·2
91	16·3	10	15·5	29	17·8
92	16·5	11	15·5	30	14·4
93	13·3	12	14·2	31	15·0
94	16·5	13	15·8	32	16·0
95	15·0	14	15·7	33	16·8
96	15·9	15	14·1	34	16·9
97	15·5	16	14·8	35	16·6
98	16·9	17	14·4	36	16·2
99	16·4	18	15·6	37	14·0
1900	14·9	19	13·9	38	18·1
01	14·5	20	14·7	39	17·5
02	16·6	21	14·3		

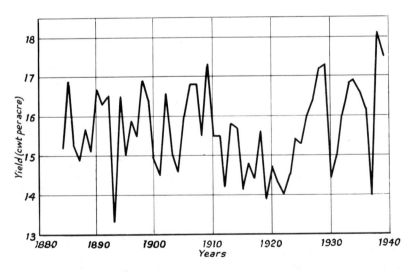

Fig. 1.3   Graph of the data of Table 1.1 (barley yields per acre)

Table 1.2   *Sheep population of England and Wales for each year from 1867 to 1939 (data from the Agricultural Statistics)*

Year	Population (10 000)	Year	Population (10 000)	Year	Population (10 000)
1867	2203	1892	2119	1917	1717
68	2360	93	1991	18	1648
69	2254	94	1859	19	1512
70	2165	95	1856	20	1338
71	2024	96	1924	21	1383
72	2078	97	1892	22	1344
73	2214	98	1916	23	1384
74	2292	99	1968	24	1484
75	2207	1900	1928	25	1597
76	2119	01	1898	26	1686
77	2119	02	1850	27	1707
78	2137	03	1841	28	1640
79	2132	04	1824	29	1611
80	1955	05	1823	30	1632
81	1785	06	1843	31	1775
82	1747	07	1880	32	1850
83	1818	08	1968	33	1809
84	1909	09	2029	34	1653
85	1958	10	1996	35	1648
86	1892	11	1933	36	1665
87	1919	12	1805	37	1627
88	1853	13	1713	38	1791
89	1868	14	1726	39	1797
90	1991	15	1752		
91	2111	16	1795		

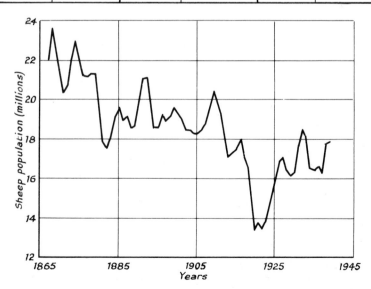

Fig. 1.4   Graph of the data of Table 1.2 (sheep population)

Table 1.3   *U.K. airlines: aircraft miles flown, by month (thousands)*

	1963	1964	1965	1966	1967	1968	1969	1970
Jan.	6 827	7 269	8 350	8 186	8 334	8 639	9 491	10 840
Feb.	6 178	6 775	7 829	7 444	7 899	8 772	8 919	10 436
Mar.	7 084	7 819	8 829	8 484	9 994	10 894	11 607	13 589
Apr.	8 162	8 371	9 948	9 864	10 078	10 455	8 852	13 402
May	8 462	9 069	10 638	10 252	10 801	11 179	12 537	13 103
June	9 644	10 248	11 253	12 282	12 950	10 588	14 759	14 933
July	10 466	11 030	11 424	11 637	12 222	10 794	13 667	14 147
Aug.	10 748	10 882	11 391	11 577	12 246	12 770	13 731	14 057
Sept.	9 963	10 333	10 665	12 417	13 281	13 812	15 110	16 234
Oct.	8 194	9 109	9 396	9 637	10 366	10 857	12 185	12 389
Nov.	6 848	7 685	7 775	8 094	8 730	9 290	10 645	11 595
Dec.	7 027	7 602	7 933	9 280	9 614	10 925	12 161	12 772

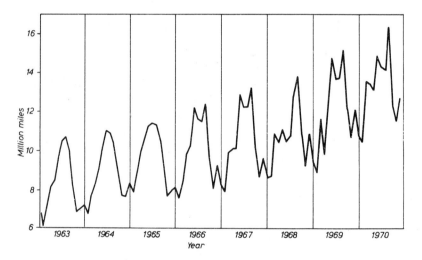

Fig. 1.5   Graph of the data of Table 1.3 (U.K. airlines: miles flown, by month)

The object of time-series analysis

1.19   It would, perhaps, be otiose at this point to enter into a minute analysis of the various reasons for which we wish to study time-series, but a
few general comments are necessary because those reasons very often determine a choice of the methods which we decide to apply. Broadly speaking,
we may delimit five types of study.

(a)   At the most superficial level we take one particular series and construct
a simple system, usually of a more or less mathematical kind, which *de-
scribes* its behaviour in a concise way.

Table 1.4 *Immigration into the United States (thousands) (Dewey, 1963)*

1820	8 385	1870	387 203	1920	430 001
1821	9 127	1871	321 350	1921	805 228
1822	6 911	1872	404 806	1922	309 556
1823	6 354	1873	459 803	1923	522 919
1824	7 912	1874	313 339	1924	706 896
1825	10 199	1875	227 498	1925	294 314
1826	10 837	1876	169 986	1926	304 488
1827	18 875	1877	141 857	1927	335 175
1828	27 382	1878	138 469	1928	307 255
1829	22 520	1879	177 826	1929	279 678
1830	23 322	1880	457 257	1930	241 700
1831	22 633	1881	669 431	1931	97 139
1832	48 386	1882	788 992	1932	35 576
1833	58 640	1883	603 322	1933	23 068
1834	65 365	1884	518 592	1934	29 470
1835	45 374	1885	395 346	1935	34 956
1836	76 242	1886	334 203	1936	36 329
1837	79 340	1887	490 109	1937	50 244
1838	38 914	1888	546 889	1938	67 895
1839	68 069	1889	444 427	1939	82 998
1840	84 066	1890	455 302	1940	70 756
1841	80 289	1891	560 319	1941	51 776
1842	104 565	1892	579 663	1942	28 781
1843	69 994	1893	439 730	1943	23 725
1844	78 615	1894	285 631	1944	28 551
1845	114 371	1895	258 536	1945	38 119
1846	154 416	1896	343 267	1946	108 721
1847	234 968	1897	230 832	1947	147 292
1848	226 527	1898	229 299	1948	170 570
1849	297 024	1899	311 715	1949	188 317
1850	295 984	1900	448 572	1950	249 187
1851	379 466	1901	487 918	1951	205 717
1852	371 603	1902	648 743	1952	265 520
1853	368 645	1903	857 046	1953	170 434
1854	427 833	1904	812 870	1954	208 177
1855	200 877	1905	1 026 499	1955	237 790
1856	200 436	1906	1 100 735	1956	321 625
1857	251 306	1907	1 285 349	1957	326 867
1858	123 126	1908	782 870	1958	253 265
1859	121 282	1909	751 786	1959	260 686
1860	153 640	1910	1 041 570	1960	265 398
1861	91 918	1911	878 587	1961	271 344
1862	91 985	1912	838 172	1962	283 763
1863	176 282	1913	1 197 892		
1864	193 418	1914	1 218 480		
1865	248 120	1915	326 700		
1866	318 568	1916	298 826		
1867	315 722	1917	295 403		
1868	277 680	1918	110 618		
1869	352 768	1919	141 132		

Note. The years are ended June 30, except for certain earlier years up to 1868, and occasional adjustments have been made to ensure comparability.

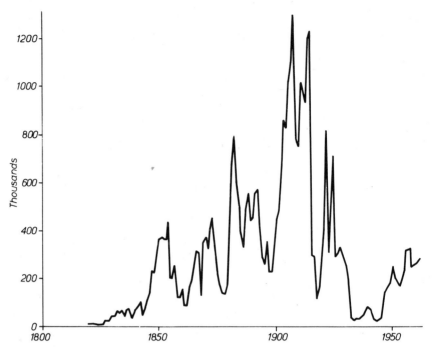

Fig. 1.6   Graph of the data of Table 1.4 (immigration into U.S.A., 1820-1962, annual)

(b) Penetrating a little deeper, we try to *explain* its behaviour in terms of other variables and to relate the observations to some structural rules of behaviour. To put it another way, we set up a model as a hypothesis to account for the observations.

(c) We may, from either (a) or (b), use the resultant analysis to *forecast* the behaviour of the series in the future. From (a) we work on the assumption that, even when we are unaware of the basic mechanism which is generating the series, there is sufficient momentum in the system to ensure that future behaviour will be like the past. From (b) we have, we hope, more insight into the underlying causation and can make projections into the future more confidently.

(d) From (b) we may require to control the system, either by throwing up warning signals of untoward events that lie ahead or by examining what might happen if we alter some of the parameters in the model.

(e) More generally, we may have to consider the joint progress through time of a number of variables, or to put it another way, our variable may be a vector of observations. In such a case we are approaching, from the

Table 1.5 *Number of normal human births in each hour in four hospital series, transformed to* y = *square root of number of births (Bliss, 1958; King, 1956)*

Hour starting	$\sqrt{\text{births}}$ = y in hospital				Total	Observed $\bar{y}$	Expected Y
	A	B	C	D			
Mt 12	13·56	19·24	20·52	21·14	74·46	18·6150	18·463
AM 1	14·39	18·68	20·37	21·14	74·58	18·6450	18·812
2	14·63	18·89	20·83	21·79	76·14	19·0350	19·129
3	14·97	20·27	21·14	22·54	78·92	19·7300	19·393
4	15·13	20·54	20·98	21·66	78·31	19·5775	19·587
5	14·25	21·38	21·77	22·32	79·72	19·9300	19·697
6	14·14	20·37	20·66	22·47	77·64	19·4100	19·716
7	13·71	19·95	21·17	20·88	75·71	18·9275	19·641
8	14·93	20·62	21·21	22·14	78·90	19·7250	19·479
9	14·21	20·86	21·68	21·86	78·61	19·6525	19·240
10	13·89	20·15	20·37	22·38	76·79	19·1975	18·941
11	13·60	19·54	20·49	20·71	74·34	18·5850	18·602
M 12	12·81	19·52	19·70	20·54	72·57	18·1425	18·246
PM 1	13·27	18·89	18·36	20·66	71·18	17·7950	17·897
2	13·15	18·41	18·87	20·32	70·75	17·6875	17·579
3	12·29	17·55	17·32	19·36	66·52	16·6300	17·315
4	12·92	18·84	18·79	20·02	70·57	17·6425	17·121
5	13·64	17·18	18·55	18·84	68·21	17·0525	17·011
6	13·04	17·20	18·19	20·40	68·83	17·2075	16·993
7	13·00	17·09	17·38	18·44	65·91	16·4775	17·067
8	12·77	18·19	18·41	20·83	70·20	17·5500	17·229
9	12·37	18·41	19·10	21·00	70·88	17·7200	17·468
10	13·45	17·58	19·49	19·57	70·09	17·5225	17·767
11	13·53	18·19	19·10	21·35	72·17	18·0425	18·106
Total	327·65	457·54	474·45	502·36	1762·00	18·3542	

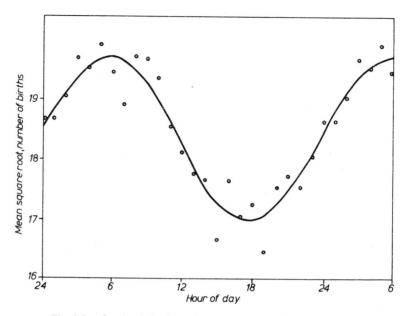

Fig. 1.7   Graph of the data of Table 1.5 (births by hour of day)

statistical angle, the more general subject of mathematical model-building, and it is sometimes hard to distinguish the point where the theory of multiple time-series ends and that of model-building begins. Perhaps there is no such point, although model-building in its most general form is a much wider subject than time-series and will not be extensively considered in this book.

Decomposition

1.20   A survey of the examples of time-series already given, and of the many others which are doubtless known to the reader, suggests that we may usefully consider the general series as a mixture of four components:

    (a) a trend, or long-term movement;

    (b) fluctuations about the trend of greater or less regularity;

    (c) a seasonal component;

    (d) a residual, irregular, or random effect.

It is convenient to represent the series as the sum of these four components, and one of the objects of analysis is to break the series down into their constituents for individual study. We must, however, remember that in doing so we are, in effect, imposing a model on the situation. It may be reasonable to suppose that trends are due to permanent forces uniformly operating in more or less the same direction, that short-term fluctuations about these long movements are due to a different set of causes, and that there is in both some

Table 1.6   *Financial Times Index of leading equity prices: quarterly averages, 1960-1971*

Year and Quarter		Index	Year and Quarter		Index	Year and Quarter		Index
1960	1	323·8	1964	1	335·1	1968	1	409·1
	2	314·1		2	344·4		2	461·1
	3	321·0		3	360·9		3	491·4
	4	312·9		4	346·5		4	490·5
1961	1	323·7	1965	1	340·6	1969	1	491·0
	2	349·3		2	340·3		2	433·0
	3	310·4		3	323·3		3	378·0
	4	295·8		4	345·6		4	382·6
1962	1	301·2	1966	1	349·3	1970	1	403·4
	2	285·8		2	359·7		2	354·7
	3	271·7		3	320·0		3	343·0
	4	283·6		4	299·9		4	345·4
1963	1	295·7	1967	1	318·5	1971	1	330·4
	2	309·3		2	343·1		2	372·8
	3	295·7		3	360·8		3	409·2
	4	342·0		4	397·8		4	427·6

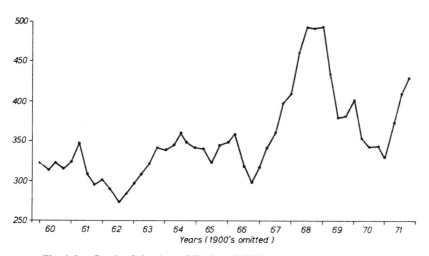

Fig. 1.8   Graph of the data of Table 1.6 (F.T. Index, quarterly averages)

disturbance attributable to random events, giving rise to the residual. But that this is so, and that the effects of the different causes are additive, is an assumption, and presupposes a type of hypothesis which we must always be ready to discard if our model fails to fit the data.

1.21   Perhaps the easiest components to understand are those which are undoubtedly due to cyclical generators, e.g. diurnal variations of temperature, the tidal movements associated with the lunar month, and seasonal variation itself. We must be careful not to confuse such effects with fluctuations of a pseudo-cyclical kind such as trade "cycles", or with sunspot "cycles" in which there is no known underlying astronomical phenomenon of a periodic kind.[*] The definition of seasonality, however, is by no means as easy as might be thought. A glance at Fig. 1.5 will illustrate one of the problems. In this series of air-miles travelled there are undoubtedly seasonal effects, a peak around Christmas, another at Easter, and one in the summer, all due to holiday travel. But the recurrence at Easter varies with Easter itself and therefore does not occur at the same date each year; and the pattern of the variation is altering from year to year, owing partly to the increased volume of traffic and partly to the spread of the period over which holidays are now taken. In short, our seasonal effect itself has a trend.

1.22   As we shall see when we come to a detailed study, it seems that trend and seasonality are essentially entangled, and we cannot isolate one without at the same time trying to isolate the other. Conceptually, however, they are distinct enough. Our general idea of trend is that of a smooth broad movement of a non-oscillatory kind extending over a considerable period of time. However, it is a relative term. What appears as a trend in climate to a drainage engineer may be nothing more than a temporary observation or a short-term swing to the geologist, whose time-scale is very much longer.

1.23   If we can determine a trend and a seasonal component and abstract them from the data we are, in general, left with a fluctuating series which may, at one extreme, be purely random, or, at the other, a smooth oscillatory movement. For the most part we have something between the two: there is some irregularity, especially in imperfect data, but also some kind of systematic effect in the sense that successive observations are not independent. We shall call this systematic effect an *oscillation*, and avoid describing it as a *cycle* unless it can be shown to be genuinely cyclical in the pattern of recurrence, and in particular that its peaks and troughs occur at equal intervals of time. Very few economic series are cyclical in this sense.

---

[*]Although it has been suggested that an apparent four-yearly swing in the British economy is due to the man-made fact that General Elections must be held at not greater than five-yearly intervals.

1.24 With a few exceptions we shall suppose that the series under study is observed at equal intervals of time, and as a rule no generality is lost if we take these intervals as units, so that we may denote a series by using subscripts, as u_1, u_2, u_3, etc., the observation at time t being u_t. Here we suppose observations to begin at $t = 1$, but if necessary we can represent previously occurring values by t_0, t_{-1}, t_{-2}, etc.

One consequence of this approach is that the mathematics of the subject is commonly expressed in terms of differences rather than of differential coefficients. The reader who is familiar with the elements of differential calculus should have no problems in extending his knowledge to the calculus of finite differences, which bears a strong formal resemblance to it. In the sequel, on the relatively rare occasions when we require a result for finite differences we shall indicate the parallel result in the differential calculus, and the reader who recognizes the latter will probably be willing to take the former on trust.

NOTES

(1) Little theoretical work has been done on time-series observed at unequal intervals. Reference may be made to Quenouille (1958) on autoregressive series and Granger (1963) on the effect of varying month-length on economic series.

(2) Daniels (1970) has studied the correlations between first differences of wandering series when, instead of cumulation or averaging over certain lengths, the mid-range of the values in a given length is taken as typical of that section of the series. (The mid-range is the mean of the largest and smallest values.) Whereas in Example 1.1 the limiting value of the correlation, as m becomes larger, is 0.25, the corresponding limit for the mid-range is even greater, namely 0.315. See also Rosenberg (1970).

(3) In theoretical statistics there also arises for study the distribution of intervals between similar events, e.g. the time-intervals between the arrival of vessels in port, or between accidents on a particular road. The study of such phenomena is usually regarded as part of the theory of stochastic processes (e.g. queuing theory or inventory control) and is not part of time-series analysis.

2

Tests of randomness

2.1 The simplest possible hypothesis that we can set up of a series which shows any chance fluctuation is that it is random. In practice a mere inspection of the data is enough to dismiss such a possibility, but there are cases where we need a more accurate test. The barley data of Table 1.1 provide a case in point. Other instances may arise when we have abstracted the systematic elements of a series and wish to test the residuals to see whether any trace of systematization remains.

2.2 In a random series, by hypothesis, the observations are independent and could have occurred in any order. There is no limit to the number of tests for randomness that we can apply, but certain criteria mark out some as better than others.

(a) If possible the test should not make any restrictive assumptions about the distribution from which the observations are supposed to arise.

(b) The calculations required should be kept to a minimum.

(c) The calculations should be easily updated, that is to say if, after having tested the series, new observations arise — as they well may with the passage of time — we should not have to perform all the calculations *ab initio*.

(d) The choice of a test depends to some extent on what alternative hypothesis we have in mind. The work of Neyman and Pearson in hypothesis testing has driven home the message that one does not test a hypothesis all by itself, but only in comparison with other possible hypotheses. We may not always be able to specify precisely what alternatives we have in mind, but we usually have some rough idea which is enough to help in deciding which of the possible tests to use. For example, if the data look as if they have a trend, we should require a different test from the case in which we suspect periodicity.

Turning points

2.3   Perhaps the easiest test to apply, especially if the series has been graphed, is to count the number of peaks or troughs which it exhibits. A "peak" is a value greater than its two neighbouring values; a "trough", conversely, is a value less than its two neighbours. The two together are known as "turning points" and the question we have to discuss is: what is the distribution of the number of turning points in a random series?

2.4   We consider, then, a finite series of n values, $u_1, u_2, \ldots, u_n$. The initial value cannot be regarded as defining a turning point because we do not know u_0; and similarly, neither can the last value since we do not know u_{n+1}. Three consecutive values are required to determine a turning point. If the series is random, these three values could have arisen in any of the six possible orders with equal probability. In only four of these would there be a turning point, namely when the greatest or the least of the three is in the middle. The probability of finding a turning point in any set of three values is then $\frac{2}{3}$.

For the set of n values, define a "counting" variable X as

$$X_i = 1, \quad u_i < u_{i+1} > u_{i+2}$$
$$\text{or } u_i > u_{i+1} < u_{i+2},$$
$$= 0 \quad \text{otherwise.} \tag{2.1}$$

The number of turning points p in the series is then simply

$$p = \sum_{i=1}^{n-2} X_i, \tag{2.2}$$

We have at once

$$\mathrm{E}(p) = \Sigma \mathrm{E}(X_i) = \tfrac{2}{3}(n-2). \tag{2.3}$$

This is the number of turning points we may expect (in other words we expect a turning point about every $1\frac{1}{2}$ observations). If there are more than this (a rare case) the series is fluctuating rapidly in a manner which cannot be due to chance alone. If there are fewer the successive values are positively correlated. We require, however, the variance p to decide when the difference between the observed and the expected number is significant. Now

$$\mathrm{E}(p^2) = \mathrm{E}\left\{\sum_{i=1}^{n-2} X_i\right\}^2$$

$$= \mathrm{E}\left\{\sum_{n-2} X_i^2 + 2\sum_{n-3} X_i X_{i+1} + 2\sum_{n-4} X_i X_{i+2} + \sum_{(n-4)(n-5)} X_i X_{i+j}\right\},$$
$$j \neq 0, 1, 2. \tag{2.4}$$

In expanding the square of ΣX we have separated those terms which are neighbours, one apart, two apart, and more than two apart. The subscripts of the Σ's denote the number of terms concerned. As a check on their accuracy we observe that

$$n - 2 + 2(n - 3) + 2(n - 4) + (n - 4)(n - 5) = (n - 2)^2 \qquad (2.5)$$

The expectations of each term in (2.4) have to be considered separately.
Since $X_i^2 = X_i$ we have immediately

$$E(X_i^2) = \tfrac{2}{3}. \qquad (2.6)$$

Also, in the last term of (2.4) X_i and X_{i+j} are independent, being at least
three terms apart. Hence

$$E(X_i X_{i+j}) = E(X_i) E(X_{i+j}) = \tfrac{4}{9}. \qquad (2.7)$$

To evaluate $E(X_i X_{i+1})$ consider a set of four consecutive terms, which in
ascending order of magnitude may be denoted by $1, 2, 3, 4$. There are $4! = 24$
ways in which these observations may occur:

1 2 3 4	2 1 3 4	3 1 2 4	4 1 2 3
1 2 4 3	2 1 4 3	3 1 4 2	4 1 3 2
1 3 2 4	2 3 1 4	3 2 1 4	4 2 1 3
1 3 4 2	2 3 4 1	3 2 4 1	4 2 3 1
1 4 2 3	2 4 1 3	3 4 1 2	4 3 1 2
1 4 3 2	2 4 3 1	3 4 2 1	4 3 2 1

$$(2.8)$$

In the first of these there are no turning points. In the second (reading across)
is a trough defined by 213 and no peak; likewise in the third; and so on. The
only non-vanishing contribution to $X_i X_{i+1}$ occurs when there are both peak
and trough in the set, e.g. in 1324 or 4231. There are, in fact, 10 such cases
among the 24 and hence

$$E(X_i X_{i+1}) = \tfrac{5}{12}. \qquad (2.9)$$

For the expectation of $X_i X_{i+2}$ we have to consider five consecutive terms.
If the reader will write down the $5! = 120$ permutations of the numbers 1 to
5 he will find that there are 54 non-vanishing contributions to $X_i X_{i+2}$. Thus

$$E(X_i X_{i+2}) = \tfrac{9}{20}. \qquad (2.10)$$

Then from equation (2.4) we have

$$E(p^2) = \tfrac{2}{3}(n-2) + \tfrac{5}{6}(n-3) + \tfrac{9}{10}(n-4) + \tfrac{4}{9}(n-4)(n-5)$$
$$= \frac{40n^2 - 144n + 131}{90}. \qquad (2.11)$$

Hence

$$\text{var } p = E(p^2) - \{E(p)\}^2 = \frac{16n - 29}{90}. \qquad (2.12)$$

By the same method, but with considerably more algebra, we can find higher
moments of p. The third and fourth cumulants are given by

$$\kappa_3(p) = -\frac{16(n+1)}{945}.$$ (2.13)

$$\kappa_4(p) = \frac{-1408n + 3317}{18\,900}.$$ (2.14)

Thus, in standard measure $\kappa_3/\kappa_2^{3/2}$ is approximately

$$-\frac{16n}{945}\bigg/\left(\frac{16n}{90}\right)^{3/2} = -0\cdot2n^{-\frac{1}{2}}$$

and $\kappa_4/\kappa_2^2 = $ approximately $-2\cdot4n^{-1}$. This indicates a fairly rapid tendency of the distribution to normality. The consequence is that we may test an observed value against the expected value in the normal distribution with a standard deviation given by $\{(16n - 29)/90\}^{\frac{1}{2}}$.

Example 2.1

In the barley data of Table 1.1 there are 56 values, but at two points (1906/7 and 1910/11) the values in successive years are equal. We shall take each of these as a single point and, to be on the safe side, reduce the number n to 54.

There are 35 turning points in the series. The expected number is $\frac{2}{3} \times 54 = 36$. Agreement is so close that no test is necessary; had it been, we should have found the variance to be $(16 \times 54 - 29)/90 = 9\cdot278$ with a standard deviation of about $3\cdot04$.

Phase-length

2.5 It is of some interest to consider not merely the numbers of turning points but the distribution of intervals between them. The interval between two turning points is called a "phase". Thus, if u_i is a trough and u_{i+1} a peak, there would be a phase of 1 between them.

To define a phase of length d (say, a run up), we require $d + 3$ terms, involving a fall from first to second, a run of rises to the $(d + 2)$th and a fall to the $(d + 3)$rd. Consider such a set of $d + 3$ values in increasing order of magnitude. Leaving the two end-points fixed, if we pick out a pair from the other $(d + 1)$ and transfer one to the beginning and the other to the end, we obtain a phase of length d. There are $\frac{1}{2}d(d + 1)$ ways of picking out the pair, and each can go to either end, so the number of rising phases is $d(d + 1)$. In addition we may put the first member at the end and any of the others except the second at the beginning, giving us $(d + 1)$ further cases; and the last member at the beginning and any except the penultimate member at the end, giving $(d + 1)$ more, except that we have counted twice the case where the first is put last and the last first. Thus there are

$$d(d + 1) + 2(d + 1) - 1 = d^2 + 3d + 1$$

rising phases. The probability that a set has either a rising or a falling phase
is then

$$\frac{2(d^2 + 3d + 1)}{(d + 3)!} \tag{2.15}$$

In a series of length n there are $n - d - 2$ possible consecutive sets of $d + 3$.
Thus the expected number of phases of length d in the whole series is

$$\frac{2(n - d - 2)(d^2 + 3d + 1)}{(d + 3)!} \tag{2.16}$$

and the total number of phases from length 1 to length $n - 3$ is N, say, where

$$N = 2 \sum_{d=1}^{n-3} \frac{(n - d - 2)(d^2 + 3d + 1)}{(d + 3)!}$$

$$= 2 \sum \left\{ -\frac{1}{d!} + \frac{n + 1}{(d + 1)!} - \frac{2n + 1}{(d + 2)!} + \frac{n + 1}{(d + 3)!} \right\}, \tag{2.17}$$

reducing to

$$N = 2 \left(\frac{2n - 7}{6} + \frac{1}{n!} \right). \tag{2.18}$$

For all practical purposes we may neglect the factor $1/n!$ so that approximately

$$N = \tfrac{1}{3}(2n - 7). \tag{2.19}$$

Example 2.2

In the barley data there are 34 phases. Their actual lengths and the theoretical values given by (2.16) are as follows:

Phase-length	No. of phases observed	Theoretical
1	23	21·25
2	7	9·17
3	4	2·59
Total	34	33·01

The agreement is close and a significance test is hardly necessary.

2.6 A comparison of observed and theoretical numbers of phases by a χ^2-test
of the usual kind is invalidated by the fact that the lengths of phases are not
independent. Wallis and Moore (1941) suggest that for a three-fold classifi-
cation with $d = 1, 2, \geqslant 3$, χ^2 can be tested with $2\tfrac{1}{2}$ degrees of freedom for
values $\geqslant 6\cdot 3$ and $\tfrac{6}{7}\chi^2$ can be tested with two degrees of freedom for lower
values.

The distribution of phase-lengths does not tend to normality for large n. The number of phases, however, does so — see Levene (1952). Gleissberg (1945) tabulated the actual distribution of this number for $n \leqslant 25$. See also Kendall and Stuart, vol. 3, pages 353-5 and 363.

The difference-sign test

2.7   A somewhat more laborious test may be conducted by counting the number of positive first differences of the series, that is to say, the number of points where it increases. With a series of n terms we have $n-1$ differences. As before, define a "counting variable"

$$X_i = 1, \quad u_{i+1} > u_i$$
$$= 0, \quad u_{i+1} < u_i. \tag{2.20}$$

For a random series we have immediately for the number of points of increase, say c,

$$E(c) = E\left(\sum_{i=1}^{n-1} X_i\right) = \tfrac{1}{2}(n-1). \tag{2.21}$$

The distribution tends fairly quickly to normality. In the foregoing manner we find

$$E(c^2) = E\left\{\sum_{n-1} X_i^2 + 2\sum_{n-2} X_i X_{i+1} + \sum_{(n-2)(n-3)} X_i X_j\right\}, \quad j \neq i, i+1,$$

and by considering permutations of 3 (the details are left to the reader as an exercise) we find

$$\text{var } c = \tfrac{1}{12}(n+1). \tag{2.22}$$

The distribution was tabulated by Moore and Wallis (1943).

2.8   The difference-sign test is clearly useless for an oscillatory series, in which the number of points of increase will in any case be approximately $\tfrac{1}{2}n$. It has been mainly advocated as a test against linear trend. On the other hand, the test based on turning points may have a poor performance as a test for trend, because the imposition of a marked random fluctuation on a mild trend will have much the same set of turning points as if the trend were absent. A better but more laborious test for linear trend is to regress u on t and to test the significance of the regression coefficient, or to use the τ coefficient described in the following paragraph.

Rank tests

2.9   We may extend the idea of comparing neighbouring values in the series to comparing all values. Given a series $u_1, \ldots, u_n$, let us count the number of cases in which $u_j > u_i$ for $j > i$. Call this number P. There are $\tfrac{1}{2}n(n-1)$ pairs for comparison and the expected number in a random series is $\tfrac{1}{4}n(n-1)$. The excess of P over this number, if significant, suggests a rising trend; a

· deficiency suggests a falling trend.

The number P, in fact, is simply related to a coefficient of rank correlation sometimes known as Kendall's τ. We define

$$\tau = \frac{4P}{n(n-1)} - 1. \tag{2.23}$$

This coefficient may vary from -1 to $+1$. Its expected value in a random series is zero, and its variance is given by

$$\text{var}\ \tau = \frac{2(2n+5)}{9n(n-1)}. \tag{2.24}$$

These results we quote without proof. (See Kendall, 1969.)

2.10 There are a number of other results in this field which have considerable theoretical interest, but they are so seldom required in practice that we will content ourselves with a brief summary and references for the reader who wishes to pursue the subject further.

(a) A test for linear trend is not often required, but when it is, the best tests are either a linear regression or the coefficient τ. The latter has the advantage that it requires no machine calculation and can be easily updated. It may be shown that the difference-sign test, as a test for trend, has zero asymptotic relative efficiency against either the regression coefficient or τ (Mann, 1945; Stuart, 1954, 1956; Kendall and Stuart, vol. 3).

(b) As a test for randomness against systematic oscillation when no trend is suspected the count of turning points is simple to apply and effective in practice. But if the turning points appear to bunch together the phase test is more refined.

(c) Foster and Stuart (1954) considered the distribution of records in a series, a record being a value which is greater than or less than all previously noted values. As a test for trend it is less efficient than the regression coefficient or τ. The main disadvantage is, of course, that as time goes on, records tend to become sparse unless the trend is fairly marked.

2.11 It was observed at the beginning of the chapter that we may wish to test for randomness a set of residuals obtained by abstracting systematic elements from a series. Unfortunately the process of abstraction itself usually generates correlation among the observed residuals, even when the real residuals are random. It is therefore rather hazardous to apply the foregoing tests to a set of observed residuals without some examination of the distortion induced by the abstracting process. We shall revert to this topic in Chapter 3 and in 12.11.

2.12   A series of random fluctuations is essentially discontinuous, but some series of a continuous kind (the edge of a razor blade under the microscope, the trace of a sound recorder in traffic) present a very unsystematic appearance. If we pursue physical phenomena down to the atomic level they are, of course, discontinuous. The question remains whether we can have a mathematically continuous random series. The answer, in my opinion, is in the negative. Nevertheless, we may consider a series in which the interval of observation is large compared with the number of points at which random effects occur. For some purposes they may, like the razor's edge, be thought of as continuous and random; but we have to be careful about rigorous mathematical arguments in such cases. It does not seem possible to proceed to the limit of a continuum in the way that mathematicians establish the arithmetic continuum from sets of discontinuously occurring points.

NOTE

All the tests given in this chapter are distribution-free, except the ordinary test of a regression coefficient when a variable is regressed on time to determine a linear trend. Most of the series encountered in practice are so obviously non-random that an elaborate discussion of tests for randomness would scarcely repay the effort involved. However, in the theory of stationary processes (for example, in Chapter 7) we often find that exact results of a distributional kind can be obtained only for random series, and such results act as a useful check on approximate formulae for non-random cases.

3

Trend

3.1 The essential idea of trend is that it shall be smooth, which in practice means that we should like to represent it by a continuous and differentiable function of the time. There may be several functions which are appropriate to particular circumstances. In the sheep series of Table 1.2, for example, it looks as if a low order polynomial might provide a reasonable account of the general downward movement of the system. For the births data of Table 1.5 harmonics are required. For the airline data of Table 1.3 a polynomial might fit the segment of the series given but would clearly be unsafe to project far into the future, because the variable must level off sooner or later and no polynomial can have a horizontal asymptote. There are occasions when we may wish to fit a functional form over the whole course of the observations, but they are rather exceptional, and the procedure has at least four practical disadvantages:

(a) It is troublesome to update.

(b) If the function contains several parameters the estimates of those parameters may be rather unreliable; for example, fitting a polynomial of order p depends on moments of order $2p$.

(c) If the fitted form is updated the whole course of the series of fitted values may be affected, and it undesirable that new observations at further points of time should require a reassessment of the past, especially the distant past.

(d) As we shall see, the dissection of seasonal and trend elements has to proceed by a kind of iterative process involving at least two estimations of each, and the resultant arithmetic would be substantial even on a fast computer.

Moving averages
3.2 The preferable procedure, since any smooth function can, under very general conditions, be represented locally by a polynomial to a fairly high

degree of accuracy, is as follows. We fit a polynomial to the first set of terms, say $2m + 1$ (an odd number, for reasons which will become apparent), and use that polynomial to determine the trend value at the $(m + 1)$th point, the middle of the range of the set. We then fit the same order of polynomial to the 2nd, 3rd, ..., $(2m + 2)$th observations and determine the trend point at the $(m + 2)$th point, and so on, working our way along the series to the last group of $(2m + 1)$. In actuality we do not have to fit the polynomials each time. As we now show, the procedure is equivalent to taking linear combinations of the observations with coefficients which can be worked out once and for all.

3.3   Suppose, for example, we wish to fit a polynomial of order three to sets of seven points. Without loss of generality we take the time points to be $t = -3, -2, -1, 0, 1, 2, 3$. Our polynomial may be written

$$u_t = a_0 + a_1 t + a_2 t^2 + a_3 t^3. \tag{3.1}$$

We determine the constants a on the principle of least squares, i.e. so as to minimize

$$\sum_{t=-3}^{3} (u_t - a_0 - a_1 t - a_2 t^2 - a_3 t^3)^2. \tag{3.2}$$

Differentiation by the a's gives us the four equations typified by

$$\sum u_t t^j - a_0 \sum t^j - a_1 \sum t^{j+1} - a_2 \sum t^{j+2} - a_3 \sum t^{j+3} = 0,$$
$$j = 0, 1, 2, 3. \tag{3.3}$$

Now the sums of the odd powers of t from -3 to $+3$ vanish and the equations reduce to

$$\sum u_t = 7a_0 \qquad\quad + 28a_2$$
$$\sum t u_t = \qquad\quad 28a_1 \qquad\quad + 196a_3$$
$$\sum t^2 u_t = 28a_0 \qquad + 196a_2$$
$$\sum t^3 u_t = \qquad\quad 196a_1 \qquad\quad + 1588a_3 \tag{3.4}$$

We are, for the present, interested only in a_0, the value of the series at $t = 0$. We then require only the first and third of these equations to find

$$a_0 = \tfrac{1}{21}\left\{ 7 \sum_{t=-3}^{3} u_t - \sum_{t=-3}^{3} t^2 u_t \right\}$$
$$= \tfrac{1}{21} \{7(u_{-3} + u_{-2} + u_{-1} + u_0 + u_1 + u_2 + u_3)$$
$$- (9u_{-3} + 4u_{-2} + u_{-1} + 0 + u_1 + 4u_2 + 9u_3)\}$$
$$= \tfrac{1}{21} \{-2u_{-3} + 3u_{-2} + 6u_{-1} + 7u_0 + 6u_1 + 3u_2 - 2u_3\}. \tag{3.5}$$

The trend value at any point is then an average of the seven points of which that point is the centre, with weights

$$\tfrac{1}{21}[-2, 3, 6, 7, 6, 3, -2] \tag{3.6}$$

which, in virtue of the symmetry, we can shorten to

$$\tfrac{1}{21}[-2, 3, 6, 7, \dots].\tag{3.7}$$

For obvious reasons this is known as a *moving average*. Consider, for example, the series

$$t = 1\ 2\ 3\ 4\ \ 5\ \ 6\ \ 7\ \ 8\ \ 9\ \ 10$$
$$\text{Series} = 0, 1, 8, 27, 64, 125, 216, 343, 512, 729$$

The trend value at $t = 4$ would be

$$\tfrac{1}{21}\{(-2 \times 0) + (3 \times 1) + (6 \times 8) + (7 \times 27) + (6 \times 64) + (3 \times 125) - (2 \times 126)\}$$
$$= 27,$$

and clearly this is correct since we are fitting a cubic to a cubic.

3.4   The procedure is perfectly general. If we fit $2m + 1$ points by a polynomial of order p, we have to minimize

$$\sum_{-m}^{m} (u_t - a_0 - a_1 t - \dots - a_p t^p)^2.$$

This leads to $p + 1$ equations analogous to (3.3) and they split into two sets as in (3.4). The solution for a_0 depends on numerical values determined by the sum $\Sigma\, t^j$ and linear functions of the u's typified by $\Sigma\, t^j u_t$. The trend value at point $t = k$ is then a linear average of the values u_{k-m} to u_{k+m}. The following are the formulae up to a moving average of 21 points and a quintic polynomial:

Quadratic and Cubic

[5] $\tfrac{1}{35}[-3, 12, \mathbf{17}]$

[7] $\tfrac{1}{21}[-2, 3, 6, \mathbf{7}]$

[9] $\tfrac{1}{231}[-21, 14, 39, 54, \mathbf{59}]$

[11] $\tfrac{1}{429}[-36, 9, 44, 69, 84, \mathbf{89}]$

[13] $\tfrac{1}{143}[-11, 0, 9, 16, 21, 24, \mathbf{25}]$

[15] $\tfrac{1}{1105}[-78, -13, 42, 87, 122, 147, 162, \mathbf{167}]$

[17] $\tfrac{1}{323}[-21, -6, 7, 18, 27, 34, 39, 42, \mathbf{43}]$

[19] $\tfrac{1}{2261}[-136, -51, 24, 89, 144, 189, 224, 249, 264, \mathbf{269}]$

[21] $\tfrac{1}{3059}[-171, -76, 9, 84, 149, 204, 249, 284, 309, 324, \mathbf{329}]$

$$(3.8)$$

Quartic and Quintic

$$[7] \quad \tfrac{1}{231}[5, -30, 75, \mathbf{131}]$$
$$[9] \quad \tfrac{1}{429}[15, -55, 30, 135, \mathbf{179}]$$
$$[11] \quad \tfrac{1}{429}[18, -45, -10, 60, 120, \mathbf{143}]$$
$$[13] \quad \tfrac{1}{2431}[110, -198, -135, 110, 390, 600, \mathbf{677}]$$
$$[15] \quad \tfrac{1}{46189}[2145, -2860, -2937, -165, 3755, 7500, 10\,125, \mathbf{11\,063}]$$
$$[17] \quad \tfrac{1}{4199}[195, -195, -260, -117, 135, 415, 660, 825, \mathbf{883}]$$
$$[19] \quad \tfrac{1}{7429}[340, -255, -420, -290, 18, 405, 790, 1110, 1320, \mathbf{1393}]$$
$$[21] \quad \tfrac{1}{260015}[11\,628, -6460, -13\,005, -11\,220, -3940, 6378,$$
$$17\,655, 28\,190, 36\,660, 42\,120, \mathbf{44\,003}]$$

$$(3.9)$$

3.5 Certain properties of such moving averages are easily derived:

(a) The weights sum to unity. This must be so because, if we apply them to a series consisting simply of the same constant repeated, the average must be that same constant.

(b) The weights are symmetric about the middle value. This is evident from the fact that they are derived from sums of terms in $\sum_{-m}^{m} t^j u_t$ which are themselves symmetric.

(c) It follows from (b) that we get the same trend values whether we fit forwards or backwards in time.

(d) As the determining equations such as (3.4) split into two groups, we get the same value for a_0 whether there is a term in $a_3 t^3$ or not. In other words, the formulae are the same for a polynomial of even order $2k$ as for the polynomial of order $2k + 1$.

We may add that

(e) as we have derived the formulae there are no trend values for the first and the last m values of the series. We shall remedy this deficiency in paragraph **3.9**.

(f) Although formulae could be derived for fitting trends to an even number of points, the result would be to give trend values half-way between the intervals of observation, which would clearly be very inconvenient. See **3.11**.

3.6 It is sometimes more convenient to express the formulae in terms of the differences of the series, especially when the differences are smaller than the original values of the series themselves. Writing for the forward difference

$$\Delta u_t = u_{t+1} - u_t, \tag{3.10}$$

so that $\qquad \Delta^2 u_t = u_{t+2} - 2u_{t+1} + u_t, \text{ etc.,}$

we have, for example,

$$\begin{aligned}
\tfrac{1}{21}[-2, 3, 6, 7, 6, 3, -2] &= \tfrac{1}{21}[-2u_{t+3} + 3u_{t+2} + 6u_{t+1} + \text{etc.}] \\
&= \tfrac{1}{21}[-2\Delta^6 u_{t-3} - 9u_{t+2} + 36u_{t+1} - 33u_t \\
&\qquad\qquad + 36u_{t-1} - 9u_{t-2}] \\
&= \tfrac{1}{21}[-2\Delta^6 u_{t-3} - 9\Delta^5 u_{t-3} - 9u_{t+1} - 57u_t \\
&\qquad\qquad - 54u_{t-1} + 36u_{t-2} - 9u_{t-3}] \\
&= \tfrac{1}{21}[-2\Delta^6 u_{t-3} - 9\Delta^5 u_{t-5} - 9\Delta^4 u_{t-3} + 21u_t] \\
&= u_t - \tfrac{1}{21}[9\Delta^4 + 9\Delta^5 + 2\Delta^6]u_{t-3}, \tag{3.11}
\end{aligned}$$

which exhibits the fact that the fitted trend line is exact if the original series is a cubic, for then differences of the fourth and higher order vanish. Equation (3.11) also has the advantage of exhibiting the residual between series and trend term explicitly.

We may also represent the moving average as an average of differences. For instance, it is easily verified that

$$\begin{aligned}
\tfrac{1}{21}[-2, 3, 6, 7, \ldots] &= u_t + \tfrac{1}{21}[2\Delta^3 u_{t-3} + 3\Delta^3 u_{t-2} - 3\Delta^3 u_{t-1} - 2\Delta^3 u_t] \\
&= u_t + \tfrac{1}{21}[2, 3, -3, -2]\Delta^3 u_{t-3} \\
&= u_t - \tfrac{1}{21}[2, 5, 2]\Delta^4 u_{t-3}. \tag{3.12}
\end{aligned}$$

The following are the formulae of type (3.12).

Quadratic and Cubic

$$\left.\begin{array}{ll}
[5] & u_3 - \tfrac{3}{35}[1]\Delta^4 u_1 \\[4pt]
[7] & u_4 - \tfrac{1}{21}[2, 5, 2]\Delta^4 u_2 \\[4pt]
[9] & u_5 - \tfrac{1}{231}[21, 70, 115, 70, 21]\Delta^4 u_3 \\[4pt]
[11] & u_6 - \tfrac{1}{429}[36, 135, 280, \mathbf{385}]\Delta^4 u_4 \\[4pt]
[13] & u_7 - \tfrac{1}{143}[11, 44, 101, 168, \mathbf{210}]\Delta^4 u_5 \\[4pt]
[15] & u_8 - \tfrac{1}{1105}[78, 325, 790, 1435, 2100, \mathbf{2478}]\Delta^4 u_6 \\[4pt]
[17] & u_9 - \tfrac{1}{323}[21, 90, 227, 434, 686, 924, \mathbf{1050}]\Delta^4 u_7 \\[4pt]
[19] & u_{10} - \tfrac{1}{2261}[136, 595, 1540, 3045, 5040, 7266, 9240, \mathbf{10\,230}]\Delta^4 u_8 \\[4pt]
[21] & u_{11} - \tfrac{1}{3059}[171, 760, 2005, 4060, 6930, 10\,416, 14\,070, 17\,160, \\[4pt]
& \qquad\qquad\qquad\qquad\qquad\qquad\qquad\qquad \mathbf{18\,645}]\Delta^4 u_9
\end{array}\right\} \tag{3.13}$$

Quartic and Quintic

$$
\begin{aligned}
&[7]\quad u_4 + \tfrac{5}{231}[1]\Delta^6 u_1 \\
&[9]\quad u_5 + \tfrac{5}{429}[3, 7]\Delta^6 u_2 \\
&[11]\quad u_6 + \tfrac{1}{429}[18, 63, \mathbf{98}]\Delta^6 u_3 \\
&[13]\quad u_7 + \tfrac{1}{2431}[110, 462, 987, \mathbf{1302}]\Delta^6 u_4 \\
&[15]\quad u_8 + \tfrac{1}{46189}[2145, 10\,010, 24\,948, 42\,273, \mathbf{51\,198}]\Delta^6 u_5 \\
&[17]\quad u_9 + \tfrac{1}{4199}[195, 975, 2665, 5148, 7623, \mathbf{8778}]\Delta^6 u_6 \\
&[19]\quad u_{10} + \tfrac{1}{7429}[340, 1785, 5190, 10\,875, 18\,018, 24\,453, \mathbf{27\,258}]\Delta^6 u_7 \\
&[21]\quad u_{11} + \tfrac{1}{260015}[11\,628, 63\,308, 192\,423, 426\,258, 759\,003, \\
&\hspace{6.5cm} 1\,135\,134, 1\,450\,449, \mathbf{1\,581\,294}]\Delta^6 u_8
\end{aligned}
\qquad (3.14)
$$

3.7   The rather cumbrous numbers occurring in these formulae, though no longer a great arithmetic inconvenience to the computor, have led in the past, and still lead, to a desire for simpler formulae, albeit of an approximative or less efficient nature. We may construct an almost infinite number of formulae for any given order of polynomial p, but there is a price to pay for simplicity.

Example 3.1

Any average of the form

$$
u_t + b\,\Delta^4 u_{t-2} + c\Delta^5 u_{t-2} + d\Delta^6 u_{t-3} \qquad (3.15)
$$

will accurately reproduce a cubic, whatever the values of b, c, d. We may re-write it as

$$
\begin{aligned}
u_t &+ b(u_{t+2} - 4u_{t+1} + 6u_t - 4u_{t-1} + u_{t-2}) \\
&+ c\{u_{t+3} - 5u_{t+2} + 10u_{t+1} - 10u_t + 5u_{t-1} - u_{t-2}\} \\
&+ d\{u_{t+3} - 6u_{t+2} - 15u_{t+1} - 20u_t + 15u_{t-1} - 6u_{t-2} + u_{t-3}\} \\
&= u_{t+3}(c + d) + u_{t+2}(b - 5c - 6d) + u_{t+1}\{-4b + 10c + 15d\} \\
&\quad + u_t(1 + 6b - 10c - 20d) + u_{t-1}(-4b + 5c + 15d) \\
&\quad + u_{t-2}(b - c - 6d) + du_{t-3}.
\end{aligned}
\qquad (3.16)
$$

Let us put the three middle coefficients equal to zero. We then have

$$
\begin{aligned}
-4b + 10c + 15d &= 0 \\
1 + 6b - 10c - 20d &= 0 \\
-4b + 5c + 15d &= 0,
\end{aligned}
$$

giving
$$
b = -\tfrac{3}{2}, \quad c = 0, \quad d = -\tfrac{2}{5}.
$$

The average then becomes

$$
\tfrac{1}{10}[-4, 9, 0, 0, 0, 9, -4]. \qquad (3.17)
$$

This is simpler than our formula

$$\tfrac{1}{21}[-2, 3, 6, 7, 6, 3, -2]. \tag{3.18}$$

Furthermore, (3.17) preserves symmetry and is still a 7-point average. What, then, has been lost?

The answer lies in the reduction of the residual error. If we apply (3.18) to a random series with variance var ϵ, the resulting series has variance

$$\frac{1}{(21)^2}\{2^2 + 3^2 + 6^2 + 7^2 + 6^2 + 3^2 + 2^2\}\text{var } \epsilon = 0\cdot333 \text{ var } \epsilon.$$

On the other hand, (3.14) has variance

$$\tfrac{1}{100}\{4^2 + 9^2 + 9^2 + 4^2\}\text{var } \epsilon = 1\cdot94 \text{ var } \epsilon.$$

Formula (3.18), which we know to minimize the sum of squares of residuals in virtue of its mode of formation, then reduces the variance to one-third, whereas (3.17) nearly doubles it. Whether this is a price which we are willing to pay for the simpler formula depends on individual circumstances.

The coefficient by which a moving average multiplies var ϵ is usually known as the *error-reducing power*. As we have just seen, some averages have an error-increasing power. Alternatively, taking the minimum $0\cdot333$ var ϵ as standard, we might say that (3.17) has efficiency $0\cdot333/1\cdot94 = 17$ per cent. efficiency in the matter of error reduction.

3.8   An alternative method of simplification is to reduce the moving average to an iteration of simple averages, namely those in which the weights are all equal. The arithmetic then reduces mainly to summation instead of multiplication. For example, if we take a simple moving average of threes and then another simple moving average of fives of the result, we have weights given by $\tfrac{1}{15}$:

$$
\begin{array}{ccccccc}
1, & 1, & 1 & & & & \\
& 1, & 1, & 1 & & & \\
& & 1, & 1, & 1 & & \\
& & & 1, & 1, & 1 & \\
& & & & 1, & 1, & 1 \\
\hline
[1, & 2, & 3, & 3, & 3, & 2, & 1].
\end{array}
$$

The shortest, but somewhat sophisticated, method of deriving formulae based on iterative simple averages is as follows.

If a second "central" difference is defined as

$$\delta^2 u_t = u_{t+1} - 2u_t + u_{t-1} \tag{3.19}$$

it may be shown that a simple moving average of k terms, which we write as $\frac{1}{k}[k]$, is given by

$$\frac{1}{k}[k]u_t = u_t + \frac{k^2-1}{2^2 3!}\delta^2 u_t + \frac{(k^2-1)(k^2-3^2)}{2^4 5!}\delta^4 u_t + \text{etc.} \quad (3.20)$$

(see Kendall and Stuart, vol. 3, chapter 46).

For instance, if our series is to be represented by a cubic, so that fourth differences vanish, we shall have exactly

$$\frac{1}{k}[k]u_t = u_t + \frac{k^2-1}{24}\delta^2 u_t. \quad (3.21)$$

Example 3.2   Spencer's 15-point formula

Consider three successive averages of 4, 4, 5 terms with equal weights. So far as second differences we have, from (3.21)

$$\frac{1}{80}[4][4][5]u_t = u_t + \tfrac{1}{24}(4^2-1+4^2-1+5^2-1)\delta^2 u_t$$
$$= u_t + \tfrac{9}{4}\delta^2 u_t.$$

Multiplying by $1-\tfrac{9}{4}\delta^2 u_t$ we then have, so far as second differences,

$$u_t = \tfrac{1}{80}[4]^2[5][1-\tfrac{9}{4}\delta^2]u_t,$$

and substituting from (3.19),

$$u_t = \tfrac{1}{320}[4]^2[5][-9, 22, -9]. \quad (3.22)$$

Without affecting the accuracy we can add any fourth or higher differences. Let us then add to the factor $[-9, 22, -9]$ a term $-3\delta^4$, which is $[-3, 12, -18, 12, -3]$. The result is

$$u_t = \tfrac{1}{320}[4]^2[5][-3, 3, 4, 3, -3]. \quad (3.23)$$

The extent of the average is 15. It consists of four repeated averages, three of them simple arithmetic means. The weights in full are

$$\tfrac{1}{320}[-3, -6, -5, 3, 21, 46, 67, 74, \ldots]. \quad (3.24)$$

This is known as Spencer's 15-point moving average, after the actuary who introduced it in 1904. It reduces var ϵ by a factor of 0.193 as compared with the 15-point formula of (3.8) which reduces it by 0.165.

Example 3.3   Spencer's 21-point formula

Spencer is also responsible for a 21-point formula, again accurate to the cubic order. As before, we find

$$\tfrac{1}{175}[5]^2[7] = 1 + 4\delta^2,$$

giving

$$u_t = \tfrac{1}{175}[5]^2[7][-4, 9, 4].$$

We now add to $[-4, 9, 4]$ the expression

$$-3\delta^4 - \tfrac{1}{2}\delta^6 = [-3, 12, -18, 12, -3] + [-\tfrac{1}{2}, 3, -7\tfrac{1}{2}, 10, -7\tfrac{1}{2}, 3, -\tfrac{1}{2}],$$

giving

$$u_t = \tfrac{1}{350}[5]^2[7][-1, 0, 1, 2, 1, 0, -1]. \tag{3.25}$$

This has an error-reduction power of about 0·128 as against the 21-point of (3.8) which has 0·108.

The full weights are

$$\tfrac{1}{350}[-1, -3, -5, -5, -2, 6, 18, 33, 47, 57, 60, \ldots]. \tag{3.26}$$

End-effects

3.9   As we have derived them, the formulae provide no trend values for the first or last m terms of the series. The absence of values at the beginning is usually of slight importance; to have values at the end is usually essential. They can be obtained without great difficulty by an extension of the method we have already used.

In fact, in section **3.3** we fitted a cubic to seven points which we may take to be the last seven points in a series. To find the values of this cubic at $t = 1, 2, 3$ (measured from u_{t-4}) we require the values of a_1, a_2, a_3 in (3.1) which, up to now, we have not needed. A straightforward solution of (3.4) yields

$$a_1 = \tfrac{1}{1512}\left\{397 \sum_{-3}^{3} tu_t - 49 \sum_{-3}^{3} t^3 u_t\right\}$$

$$a_2 = \tfrac{1}{84}\left\{-4 \sum_{-3}^{3} tu_t + \sum_{-3}^{3} t^3 u_t\right\}$$

$$a_3 = \tfrac{1}{216}\left\{-7 \sum_{-3}^{3} tu_t + \sum_{-3}^{3} t^3 u_t\right\}. \tag{3.27}$$

Expressing these as moving averages of the last seven terms, we have, in an obvious notation,

$$u_t = \tfrac{1}{21}[-2, 3, 6, 7, 6, 3, -2] + \tfrac{1}{252}[-22, -67, -58, 0, 58, 67, -22]t$$
$$+ \tfrac{1}{84}[5, 0, -3, -4, -3, 0, 5]t^2 + \tfrac{1}{36}[-1, 1, 1, 0, -1, -1, 1]t^3. \tag{3.28}$$

For example, with $t = 1, 2, 3$, these reduce to

$$u_1 = \tfrac{1}{42}[1, -4, 2, 12, 19, 16, -4] \tag{3.29}$$

$$u_2 = \tfrac{1}{42}[4, -7, -4, 6, 16, 19, 8] \tag{3.30}$$

$$u_3 = \tfrac{1}{42}[-2, 4, 1, -4, -4, 8, 39]. \tag{3.31}$$

If the last seven terms were 0, 1, 8, 27, 64, 125, 216 we should have for the third from the end

$$u_1 = \tfrac{1}{42}[(1 \times 0) + (-4 \times 1) + (2 \times 8) + \ldots + (4 \times 216)] = 64.$$

The coefficients sum to unity, as they must, but they are no longer symmetrical. Moreover, since there are different formulae for different terms, a complete set of coefficients corresponding to (3.8) and (3.9) occupy rather a lot of space. They are tabulated for $p \leqslant 5$, $n \leqslant 25$ by Cowden (1962) with whose permission we quote them in Appendix A. It is to be noted that we can no longer use the same formulae for polynomials of order $2m$ as for those of order $2m + 1$.

3.10   One consequence of the fact that, as we get nearer to the end of the series, the coefficients become more and more unequal is that their sum of squares tends to increase. For example, the sum of squares of the coefficients of (3.29) to (3.31) are 0·4524, 0·4524, 0·9286, against the central value 0·3333. This as we might expect: the nearer the tails, the less reliable is the trend point, as measured by the error-reducing power at that point. The fitted curve, it has been said, tends to wag its tail.

Centred averages

3.11   We have noticed that it is most convenient to use an odd number of points for the extent of an average. But it so happens that many of the time-spans over which we wish to average comprise an even number of points – the twenty-four hours of the day, the four weeks of the month, the four quarters and twelve months of the year. It is highly desirable in such cases to bring the points of trend determination into line with the time-points of the observations. This is usually done by taking a simple arithmetic mean of two trend values.

For example, suppose we have observations on the last day of each month, say January through December. A simple moving average of 12, with weights $\frac{1}{12}[12]$, would give us a trend value in the middle of July. We therefore take the mean of the values at the middle of July and middle of August to provide a trend value at the end of July for comparison with the observed value at that point. This is easily seen to be equivalent to taking a thirteen-month average with weights

$$\frac{1}{24}[1, 2, 2, \ldots, 2, 2, 1]. \tag{3.32}$$

There are two Januaries (and in general two identical months) in the average, but each has only half the weight of the other months.

The effect of moving averages on other constituents

3.12   As yet no advice has been offered as to the best choice of the numbers $2m + 1$ and p, the extent of the average and the degree of the polynomial embodied in it. Before we can take up these points we must consider the effect of moving averages on the other constituents in an additive model.

Consider first of all what happens if we apply a simple moving average of k to a series consisting of a harmonic term $\sin (\lambda t + \alpha)$. In virtue of the sum

$$\sum_{t=1}^{k} \sin (\lambda t + \alpha) = \frac{\sin \frac{1}{2}k\lambda}{\sin \frac{1}{2}\lambda} \sin \{\tfrac{1}{2}(k + 1)\lambda + \alpha\} \tag{3.33}$$

$$\frac{1}{k}[k] \sin (\lambda t + \alpha) = \frac{\sin \frac{1}{2}k\lambda}{k \sin \frac{1}{2}\lambda} \sin \{\tfrac{1}{2}(k + 1)\lambda + \alpha\}. \tag{3.34}$$

If we take this as the trend value at the middle of the range, $\frac{1}{2}(k + 1)$, the effect of the average is then to reproduce the harmonic, with the same phase but an amplitude multiplied by $\sin \frac{1}{2}k\lambda/k \sin \frac{1}{2}\lambda$. This factor will only exceptionally be equal to unity. If λ is small it is nearly so, since $\sin \theta = \theta$ for small θ. In other cases it is substantially reduced by the factor k. Thus, if this trend value is subtracted from the original, the harmonic will be little affected if λ is small but may be nearly obliterated if λ is large, or even increased if $\sin \frac{1}{2}k\lambda$ is negative. This, on reflection, is understandable. If λ is small, the sine wave is long compared to the interval of observation and is treated as a trend. If λ is large, the wave repeats itself several times in the range of the average, which therefore comes close to zero.

3.13   For series which consist of, or can be represented as, the sum of a number of harmonics it will be clear that a simple moving average will, after abstraction of trend, emphasize the shorter oscillations at the expense of the longer ones. We may expect something of the same kind for moving averages which are not simple arithmetic means, especially those which can be approximated to by iterated simple averages. Thus there is danger that cyclical movements may be distorted by trend-abstraction. Indeed, for oscillations which are not cyclical in the strict sense the same kind of danger exists. A long swing, even if irregular, may, as it were, be mistaken for trend by the moving average and included in the trend; so that the remainder when trend is removed has lost some of the movement which we should, perhaps, wish to regard as fluctuation about the trend.

Autocorrelation

3.14   Consider now the effect of a moving average on a series of random residuals which might be present in the original series. We have already seen that the averaged series is altered in variance, usually by reduction. What is equally important is that its members are no longer independent.

   This is a convenient point to introduce a concept of fundamental importance in the analysis of time-series, that of autocorrelation. We have already touched on it but must now consider it rather more closely. Given the values $u_1, u_2, \ldots, u_n$, the $(n - 1)$ pairs $(u_1, u_2), (u_2, u_3), \ldots, (u_{n-1}, u_n)$ constitute a set of bivariate values which have a correlation coefficient of the standard kind. Likewise the $(n - 2)$ pairs $(u_1, u_3), (u_2, u_4), (u_3, u_5)$ and so on. We call the coefficient $(k-1)$ terms apart, i.e. of u_t and u_{t+k}, the "serial correlation of order k" and denote it by r_k. If the series is regarded as a parent form which is infinitely long, we refer to the autocorrelations and denote them by ρ_k. This accords with the usual convention in statistics of denoting parent values by Greek words or letters and sample values by roman type.

In full generality the serial correlation is given by

$$r_k = \frac{\dfrac{1}{n-k}\sum_{i=1}^{n-k}\left(u_i - \dfrac{1}{n-k}\sum_{i=1}^{n-k}u_i\right)\left(u_{i+k} - \dfrac{1}{n-k}\sum_{i=1}^{n-k}u_{i+k}\right)}{\left[\dfrac{1}{n-k}\sum_{i=1}^{n-k}\left\{u_i - \dfrac{1}{n-k}\sum_{i=1}^{n-k}u_i\right\}^2 \dfrac{1}{n-k}\sum_{i=1}^{n-k}\left\{u_{i+k} - \dfrac{1}{n-k}\sum_{i=1}^{n-k}u_{i+k}\right\}^2\right]^{\frac{1}{2}}}. \tag{3.35}$$

The clumsiness of the expression results from the fact that the kth order coefficient depends on the values $u_1, \ldots, u_{n-k}$ for one member of the pair and $u_{1+k}, \ldots, u_n$ for the other. For most purposes we can modify it so as to measure all the variables about the mean of the whole series of $u_1, u_2, \ldots, u_n$, and also replace the variance terms in the denominator by the variance of the whole series. The definition then simplifies greatly to

$$r_k = \frac{\dfrac{1}{n-k}\sum_{i=1}^{n-k}(u_i - \bar{u})(u_{i+k} - \bar{u})}{\dfrac{1}{n}\sum_{i=1}^{n}(u_i - \bar{u})^2} \tag{3.36}$$

where $\bar{u}$ is the mean $\sum_{i=1}^{n} u_i/n$.

Since $r_k = r_{-k}$ we shall not usually have to consider negative values of k.

3.15   The array of coefficients $\rho_1, \rho_2, \ldots$ or their sample counterparts tell us a great deal about the internal structure of the series, and their totality, graphed with k as abscissa and ρ_k as ordinate, is called the *correlogram*. As we shall see, it provides a useful diagnostic to distinguish between different types of series.

The Slutzky–Yule effect

3.16   Returning now to the moving average of random series, let the weights of the average of $2m + 1$ terms be $a_1, a_2, \ldots, a_{2m+1}$ and let ϵ be the series with zero mean. The "trend" value determined by the average is then u_t, say, given by

$$u_{t+m+1} = \sum_{j=1}^{2m+1} a_j \epsilon_{t+j}. \tag{3.37}$$

As we know,

$$E(u_t) = 0, \tag{3.38}$$

$$\text{var } u_t = \text{var } \epsilon \sum_{j=1}^{2m+1} a_j^2. \tag{3.39}$$

Now

$$\text{cov}(u_{t+m+1}, u_{t+m+1+k}) = \text{E}\{a_1\epsilon_{t+1} + a_2\epsilon_{t+2} + \ldots\}\{a_1\epsilon_{t+1+k} + \ldots\}$$

and since the ϵ's are independent this reduces to

$$\Sigma (a_1a_{1+k} + a_2a_{2+k} + \ldots + a_{2m-k+1}a_{2m+1}).$$

Hence the kth autocorrelation of the series is given by

$$\rho_k = \frac{\sum_{j=1}^{2m+1-k} a_j a_{j+k}}{\sum_{j=1}^{2m+1} a_j^2}. \tag{3.40}$$

The general consequence of this is that the derived series possesses non-vanishing autocorrelations up to order k. Moreover, for the kind of moving average we employ in practice, ρ_1 will be positive and may be quite high. Thus the derived series will be smoother than the original random series and may present the appearance of a systematic oscillation. This is known as the Slutzky–Yule effect.

Example 3.4

The following Table 3.1 shows the autocorrelations generated in a random series by the Spencer 21-point formula whose weights are given at (3.26). The correlogram is graphed in Fig. 3.1. From $k = 13$ the autocorrelations are small and from $k = 21$ onwards they vanish identically. For lower values of k they are substantial and we may expect, then, that application of the formula to a random series will smooth it a great deal. A typical series is set out in Table 3.2, based on a "rectangular" random variable, i.e. one in which each integral value from 0 to 99 can occur equally frequently. (This involves a mean of 49·5 for ϵ, but the smoothing effect, of course, is the same.)

3.17   The realization that apparently systematic fluctuations can be generated merely as the average of random events came as something of a shock when Slutzky (1927) and Yule (1927) first called attention to the fact, especially as Slutzky was able to mimic an actual trade "cycle" of the nineteenth century very closely by a moving-average process. The movements generated in this way, in fact, are much more like the kind of series observed in economics and in some of the physical sciences (e.g. meteorology) than the sums of harmonic terms. Characteristically, they exhibit varying intervals between peaks and troughs, so that instead of a strict "period" there is a distribution of intervals. Likewise the amplitudes of the movements show substantial variations from one peak to the next. It is therefore of some interest to study the distribution of intervals between successive turning points.

Table 3.1 *Correlations generated by a Spencer 21-point average*

k	$\Sigma\, a_j\, a_{j+k}$	r_k	k	$\Sigma\, a_j\, a_{j+k}$	r_k
0	17 542	1·000	11	−930	−0·053
1	16 786	0·957	12	−528	−0·030
2	14 667	0·836	13	−214	−0·012
3	11 584	0·660	14	− 27	−0·002
4	8 085	0·461	15	50	0·003
5	4 726	0·269	16	59	0·003
6	1 951	0·111	17	40	0·002
7	6	0·000	18	19	0·001
8	−1 074	−0·061	19	6	0·000
9	−1 430	−0·082	20	1	0·000
10	−1 298	−0·074	21	0	0·000

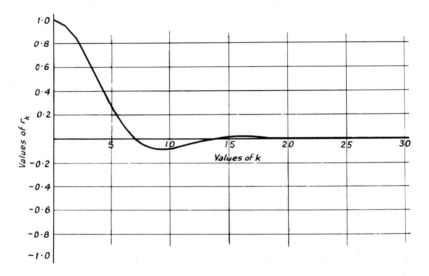

Fig. 3.1 Correlogram of the series generated by the Spencer 21-point formula

3.18 The problem of determining the actual distribution generated on a random series by a moving average has not been solved, but we may ascertain the mean length between peaks in certain cases. Let the weights of the average be $a_1, a_2, \ldots, a_s$ and apply them to a random *normal* series with zero mean and variance σ^2. Then the joint distribution of the variables $\epsilon_0, \epsilon_1, \ldots, \epsilon_{s+1}$ is a spherically symmetric function

$$dF = \frac{1}{(2\pi)^{\frac{1}{2}(s+2)}\sigma^{s+2}} \exp\left\{-\tfrac{1}{2} \sum_{j=0}^{s+1} \epsilon_j^2/\sigma^2 \sum \epsilon_j^2/\sigma^2\right\} d\epsilon_0, d\epsilon_1, \ldots, d\epsilon_{s+1}. (3.41)$$

The probability of a peak, that is to say that

$$u_{s+1} - u_s < 0 \quad \text{and} \quad u_s - u_{s-1} > 0,$$

Table 3.2 *Spencer smoothing of a rectangular random series*

Number of term	Series	Spencer 21-point "trend"	Number of term	Series	Spencer 21-point "trend"	Number of term	Series	Spencer 21-point "trend"
1	23		18	3	43	35	10	39
2	15		19	67	40	36	96	38
3	75		20	44	39	37	22	37
4	48		21	5	39	38	13	36
5	59		22	54	39	39	43	35
6	1		23	55	40	40	14	34
7	83		24	50	41	41	87	34
8	72		25	43	42	42	16	
9	59		26	10	43	43	3	
10	93		27	74	44	44	50	
11	76	67	28	35	44	45	32	
12	24	66	29	8	45	46	40	
13	97	63	30	90	44	47	43	
14	8	60	31	61	44	48	62	
15	86	55	32	18	43	49	23	
16	95	51	33	37	42	50	50	
17	23	47	34	44	41	51	5	

is the probability density between the hyperplanes

$$\sum_{j=1}^{s} a_j \, \epsilon_{j+1} - \sum_{j=1}^{s} a_j \, \epsilon_j = 0 \tag{3.42}$$

$$\sum_{j=1}^{s} a_j \, \epsilon_j - \sum_{j=1}^{s} a_j \, \epsilon_{j-1} = 0. \tag{3.43}$$

This, owing to spherical symmetry, is simply $\theta/2\pi$, where θ is the angle between the planes, given by

$$\cos \theta = \frac{(a_2 - a_1)a_1 + (a_3 - a_2)(a_2 - a_1) + \ldots + a_s(a_s - a_{s-1})}{a_1^2 + (a_2 - a_1)^2 + \ldots + a_s^2}. \tag{3.44}$$

Suppose, for example, that we generate a series by two simple averages $\frac{1}{15}[5][3]$. The weights are $[1, 2, 3, 3, 3, 2, 1]/15$ and from (3.44) we find

$$\cos \theta = \tfrac{2}{3}, \quad \theta = 48{\cdot}190.$$

The mean distance between peaks is $360/48{\cdot}190 = 7{\cdot}5$ units.

3.19   Kendall and Stuart (vol. 3, chapter 46) give an experimental verification of this case, together with formulae enabling the calculation of the mean distance between peaks when "ripples" are excluded, and of the mean distance between "upcrosses", i.e. the points where a series measured about its mean changes sign from negative to positive. Exact results for non-normal variation are not known, but it appears from rather limited experimental evidence that (3.44) is fairly robust under departures from normality.

3.20   Reverting now to the effect on random residuals of subtracting a trend from the original series, we see that the moving average will determine a "trend" in the residuals and consequently will remove part of them from the remainder after subtraction of the trend effect. This may not be very serious, because if, for example, the error-reducing coefficient is a fairly low value (e.g. about one-ninth for the Spencer 21-point) about eight-ninths of the variation will remain.

3.21   The upshot is that any moving average is likely to distort the cyclical, short-term and random effects in a series. There seems to be no escape from this situation, at least so far as trend-elimination by moving averages is concerned. Fortunately we can, knowing the weights of the average, estimate what the effect is likely to be and make some allowance for it in interpretation. We shall see shortly, however, that in any case the choice of a moving average is to some extent a matter of subjective judgement and that in some cases, at least, it is desirable to try several averages to see which of the results most nearly corresponds to the object of the analysis.

3.22   Apart from the fitting of mathematical functions over the whole course of the series, the only other method of trend determination which seems to have been attempted is that due to Rhodes (1921) and Quenouille (1949). It proceeds by cutting up the series into segments and fitting a polynomial to each. The calculations are involved, because of the necessity for ensuring that each polynomial joins smoothly on to the next, and the method has not come into general use. An account is given in Kendall and Stuart (vol. 3, chapter 46).

NOTES

(1) The effect of moving averages on constituents of a time-series may also be studied from the point of view of spectrum analysis. Cf. **6.34** and **8.27**.

(2) In certain contexts a moving average is also known as a "filter" and one encounters expressions such as a "low-pass filter", meaning, in regard to time-series, a moving average which removes constituent periodic elements of high frequency (short wavelength) and leaves relatively untouched those of low frequency (long wavelength). Cf. **6.34**.

(3) A small, but not unimportant, point for the practitioner who has to explain the details of his analysis to a layman concerns the signs of the weights in a moving average. It is not always easy to justify the use of *negative* weights, which appear to imply that some terms are deleterious to the total. It is, of course, possible to design moving averages with non-negative weights, but in general they will not be optimal, although perhaps near enough to make little difference to the trend.

(4) The tables of Appendix A contain, under the column headed "0", the result of extrapolating the polynomial one unit beyond the range of

observation and hence serve to "predict" one unit ahead. The standard errors of estimate in the final row show how the unreliability increases as we move out of the observational range.

(5) Corresponding to (3.44), it is also possible to derive expressions for the mean distance between upcrosses. From some points of view this may be a preferable measure of the average oscillation because there may occur neighbouring peaks of a ripple-like character without the series changing sign.

4

The choice of a moving average

The variate difference method

4.1   If we have a series consisting of a polynomial (or represented locally by a series of polynomials) together with a superposed random element, it is natural to consider whether we can get rid of the polynomial part by successive differencing of the series. We know, in fact, that the differences of a polynomial of order k are represented by a polynomial of order $k - 1$. If then the series embodies a polynomial of order p, successive differencing $p + 1$ times will obliterate it and leave us with elements based on the random constituent of the original series. Let us then consider the effect of differencing on a random series.

Writing, as at equation (3.10), Δ for the forward difference, we have

$$\Delta \epsilon_t = \epsilon_{t+1} - \epsilon_t \tag{4.1}$$
$$\Delta^2 \epsilon_t = \epsilon_{t+2} - 2\epsilon_{t+1} + \epsilon_t \tag{4.2}$$

and generally

$$\Delta^r \epsilon_t = \epsilon_{t+r} - \binom{r}{1}\epsilon_{t+r-1} + \binom{r}{2}\epsilon_{t+r-2} + \ldots \pm (-1)^r \epsilon_t. \tag{4.3}$$

The coefficients here are the terms in the binomial expansion of $(1 - 1)^r$.

Without loss of generality we may take ϵ to have zero mean. Then

$$E(\Delta^r \epsilon_t) = 0. \tag{4.4}$$

If ϵ_t has the same variance var ϵ for all t, we have

$$\text{var}(\Delta^r \epsilon_t) = \text{var } \epsilon \left\{ 1 + \binom{r}{1}^2 + \binom{r}{2}^2 + \ldots + 1 \right\}.$$

The term in brackets is the coefficient of x^r in $(1 + x)^r(x + 1)^r$ and is therefore $\binom{2r}{r}$. Hence

$$\text{var}(\Delta^r \epsilon_t) = \binom{2r}{r} \text{var } \epsilon. \tag{4.5}$$

We define

$$\frac{\text{var}\,(\Delta^r \epsilon_t)}{\binom{2r}{r}} \quad \text{as} \quad V_r.$$

4.2   If we then take the rth differences, sum their squares about zero and divide by $\binom{2r}{r}$ we have an estimate of var ϵ, provided that the differencing has been carried far enough to eliminate the polynomial. The variate-difference method then consists of finding first, second, third ... differences, ascertaining the sums of squares, dividing by $\binom{2}{1}, \binom{4}{2}, \binom{6}{3}$ and so on, and observing at what point the quotient settles down to a constant. At that point we have an estimate of the degree of polynomial in the original series and of the variance of the random element.

4.3   The factor $\binom{2r}{r}$ has the following values for $r = 1$ to 10:

r	$\binom{2r}{r}$	r	$\binom{2r}{r}$
1	2	6	924
2	6	7	3 432
3	20	8	12 870
4	70	9	48 260
5	252	10	184 756

4.4   The idea is attractively simple, but its practical application requires some care. The successive values of V_r are not independent, and often show a tendency to creep slowly down (and sometimes up) without ostensibly converging to a constant value. Also, the process of differencing tends to reduce the relative importance of any systematic movement, except cyclical effects with a period close to the time-interval, so that the convergence of the ratio V_r does not *prove* that the series originally consisted of a polynomial plus a random residual; only that it could be approximately represented in that way. One thing, however, the method will do for us, and that is to provide an upper limit to the order of polynomial p which can usefully be applied to eliminate trend.

Example 4.1

A series of 51 terms was constructed according to the formula

$$u_t = (t-26) + \tfrac{1}{10}(t-26)^2 + \tfrac{1}{100}(t-26)^3 + \epsilon_t, \tag{4.6}$$

where the values of ϵ_t are those set out in Example 3.4, Table 3.2. An application of the variate-difference method to the result gave the following:

r	V_r
1	1075·41
2	1082·02
3	1076·58
4	1047·21
5	1011·05
6	975·20

It appears that, so far as the method is concerned, the polynomial is of order one and the residual variance is 1075. In actual fact the polynomial is of order 3 and the residual variance (the variance of the first 100 natural numbers) is $(100^2 - 1)/12 = 833$. However, the series is too short to show up the quadratic and cubic terms. The variance of the original series was 6273. The first difference reduces it to 1075·4 and indicates, quite correctly, that about five-sixths of the variance is accounted for by the linear term. The remainder is caught up in the random term. Consider, for example, the cubic term $(t-26)^3/100$. In the original series this could range over $-156\!\cdot\!25$ to $156\!\cdot\!25$. First differences reduce it to $3(t-26)^2/100$, with a range of 18·75 to zero, whereas the random element has its range increased to 0 to 198. Already at the first difference the systematic part is being swamped by the random element.

Example 4.2

Consider a short-term periodic series consisting of a repetition of 1, -1, 1, -1. The first differences give us the series 2, -2, 2, -2, etc., the second differences the series 4, -4, 4, -4, and so on. Thus, neglecting the shortness of the series, we have

$$V_r = 2^{2r} \Big/ \binom{2r}{r} = \frac{2^{2r}(r!)^2}{(2r)!}$$

which, with Stirling's approximation to the factorial, reduces to approximately $(\pi r)^{\frac{1}{2}}$ and increases without limit. This is admittedly an extreme case. Consider then the sine series $u_t = \sin \alpha t$:

$$\Delta u_t = \sin \alpha (t+1) - \sin \alpha t = 2 \sin \tfrac{1}{2}\alpha \cos \alpha (t+\tfrac{1}{2})$$

$$\Delta^2 u_t = 2^2 \sin^2 \tfrac{1}{2}\alpha \sin \alpha (t+1) \tag{4.7}$$

$$\Delta^r u_t = 2^r \sin^r \tfrac{1}{2}\alpha \, \tfrac{\sin}{\cos} \{\alpha (t+\tfrac{1}{2}r)\}. \tag{4.8}$$

Apart from the term in $\alpha(t+\tfrac{1}{2}r)$, which has the same period and amplitude as the original, we then have

$$V_r = \frac{2^{2r} \sin^{2r} \tfrac{1}{2}\alpha (r!)^2}{(2r)!}$$

$$\sim \sin^{2r} \tfrac{1}{2}\alpha \sqrt{(\pi r)}. \tag{4.9}$$

For $\alpha = \pi$, as we have seen, this increases without limit. For other values where $|\sin \tfrac{1}{2}\alpha| < 1$ the quotient of (4.9) will diminish indefinitely, but not, perhaps, for a number of differences if the wavelength of the series is short.

Example 4.3

From the data of Table 1.2 (sheep population) we obtain the following:

r	V_r
1	3468
2	1442
3	854
4	629
5	518
6	448
7	401
8	371
9	357
10	347

This suggests that a cubic is enough to admit into trend elimination. The extra reduction by taking higher orders is slight. In point of fact, as we shall see presently, even a cubic is too high for the study of short-term oscillatory movements.

Example 4.4 (Kendall, 1946)

Although it anticipates the treatment of certain kinds of stationary series which we shall discuss fully in Chapter 6, it is worth while mentioning here the effect of variate differences on the so-called autoregressive series. These are series generated, in effect, as linear averages of random terms, but the averages are of infinite extent. They behave rather like the moving averages of random terms which we have already considered.

A number of such series were constructed according to the recurrence formula

$$u_{t+2} = -\alpha_1 u_{t+1} - \alpha_2 u_t + \epsilon_t, \qquad (4.10)$$

where, as usual, ϵ is a random term. For $\alpha_1 = -1 \cdot 1$, $\alpha_2 = 0 \cdot 5$ the variance of u can be shown to be

$$\text{var } u = 2 \cdot 8846 \text{ var } \epsilon. \qquad (4.11)$$

In this case the values of V_r can be worked out theoretically. They are as follows, standardized by division by $2 \cdot 8846$:

1	0·2657	11	0·0571
2	·1245	12	·0565
3	·0869	13	·0561
4	·0734	14	·0557
5	·0671	15	·0554
6	·0635	16	·0551
7	·0613	17	·0548
8	·0598	18	·0546
9	·0586	19	·0544
10	·0578	20	·0543

The slow downward creep is typical, and was borne out by experiment. We might well have concluded that the series could be represented by a cubic plus a random residual of variance about 0·0869 that of the original. In practice, perhaps, a moving average based on a cubic would give residuals of this order which it would be hard to disentangle from residuals actually generated on random series; but clearly we are far from the true generating mechanism of the data.

4.5 There is an interesting connection between the variances of the differences of the series and its autocorrelations. For a series of n terms, we have

$$\sum_{t=1}^{n-1} (\Delta u_t)^2 = \sum_{1}^{n-1} (u_{t+1} - u_t)^2 = \sum_{1}^{n-1} \{(u_{t+1} - \bar{u}) - (u_t - \bar{u})\}^2$$

$$= \sum_{1}^{n-1} (u_{t+1} - \bar{u})^2 + \sum_{1}^{n-1} (u_t - \bar{u})^2 - 2 \sum_{1}^{n-1} (u_{t+1} - \bar{u})(u_t - \bar{u}). \quad (4.12)$$

Neglecting end-effects, we may then write, on division by $(n-1)$,

$$\text{var } \Delta u_t = 2 \text{ var } u(1 - \rho_1). \qquad (4.13)$$

Likewise we find

$$\text{var } \Delta^2 u_t = \text{var } u\{6 - 8\rho_1 + 2\rho_2\} \qquad (4.14)$$

and generally

$$\text{var } (\Delta^r u_t) = \text{var } u\left\{\binom{2r}{r} - 2\rho_1 \binom{2r}{r-1} + 2\rho_2 \binom{2r}{r-2} \dots\right\} \qquad (4.15)$$

$$V_r = \operatorname{var} u \left\{ 1 - \frac{2r}{r+1}\rho_1 + \frac{2r(2r-1)}{(r+1)(r+2)}\rho_2 - \ldots \right\}. \qquad (4.16)$$

It may be shown conversely that

$$V_0\rho_k = V_0 - k^2 V_1 + \frac{k^2(k^2-1)}{(2!)^2} V_2 - \frac{k^2(k^2-1)(k^2-2^2)}{(3!)^2} V_3 - \ldots. \qquad (4.17)$$

Thus the variate-difference quantities V are expressible as linear functions of the autocorrelations and vice versa. The relations are rarely needed in practice, which is why we refer the reader for proofs to Kendall and Stuart (vol. 3, chapter 46).

4.6   We have remarked that the successive values of V are correlated, and attempts have therefore been made to provide standard-error type formulae to see whether V_{k+1} is "significantly" less than V_k. The results are extremely cumbrous and, in my opinion, rarely worth applying even by approximation. They are due to Oskar Anderson senior and are summarised in Kendall and Stuart (vol. 3, chapter 46).

4.7   It must be confessed that the variate-difference method is somewhat disappointing. The ideas underlying it are so clear that one would hope to obtain by its use equally clear indications of how to fit trends and estimate the residual variances. As we have exemplified, this is in general not so. However, we should not discard it as useless. It provides some valuable suggestions as to how far we should go in the order of polynomials which we use. We conclude this chapter with a discussion of a practical case, the sheep series of Table 1.2.

Example 4.5

    The sheep series, as may be seen from Fig. 1.4, shows a generally declining trend, some oscillation about that trend, and a certain amount of casual variation. Being an annual series it does not contain a seasonal element. The object of the study in this case was to abstract the trend element in order to study the oscillations, there being an impression that livestock of various kinds (pigs, sheep, cattle) were subject to some kind of cycle. From the data of Example 4.3 we conclude that a moving average reproducing a cubic is enough. We begin, then, with a Spencer 15-point.

    Fig. 4.1 shows the fitted Spencer 15-point average. It is at once clear that if we are concerned with the short-term oscillations this is much too good. The fitted curve follows those movements quite closely, treating them as "trend", and to subtract it from the primary series would remove from notice most of the variation which interests us.

    We therefore consider a simpler average. Fig. 4.2 shows the result of fitting a simple 11-point. This is obviously much closer to what we require. The trend line is not ideally smooth, but it leaves behind, so to speak, most of the short-term variation.

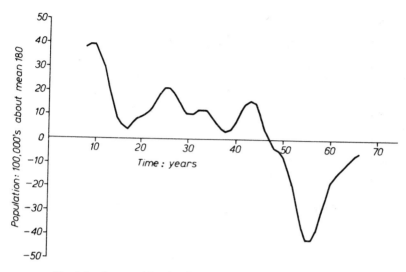

Fig. 4.1   Spencer 15-point fitted to the sheep series of Fig. 1.4

It might be supposed that an even simpler average would do even better, so Fig. 4.3 shows a simple moving average of 5. This again follows the data too closely. In the ultimate a simple average of 9 was chosen. We shall consider the residuals later in Chapter 12.

4.8   The foregoing example is enough to show that trend-fitting and trend-estimation are very far from being a purely mechanical process which can be handed over regardless to an electronic computer. In the choice of the extent of the average, the nature of the weights, and the order of the polynomial on which those weights are based, there is great scope — even a necessity — for personal judgement. To a scientist it is always felt as a departure from correctness to incorporate subjective elements into his work. The student of time-series cannot be a purist in that sense. What he can do, of course, is to make available the primary data on which he worked and explain unambiguously how he has treated them; anyone who disagrees with what has been done can then carry out his own analysis.

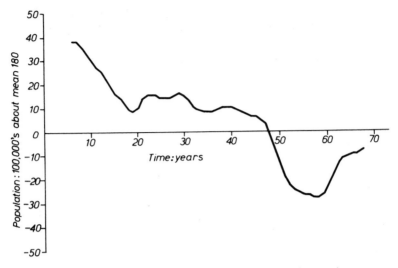

Fig. 4.2   Moving average of 11 fitted to the sheep series

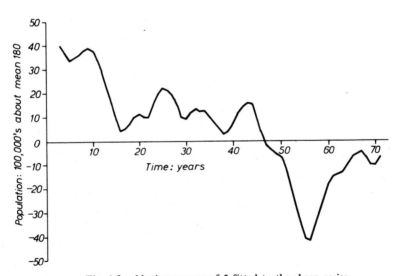

Fig. 4.3   Moving average of 5 fitted to the sheep series

5

Seasonality

5.1   Seasonal effects, although they may vary somewhat from their average time of occurrence during the year, have a degree of regularity which other elements of time-series do not, as a rule, possess. It may be possible, as we shall see when we discuss spectrum analysis, to isolate their contribution to the spectrum without regard to other constituents of the series. But for decomposition of the kind we have been considering up to this point it seems impossible to determine the seasonal effects without some prior attention to trend. Consider, for example, a monthly series consisting of a slowly rising trend, 100 for January 1970, 101 for February 1970, . . . , 112 for December 1970, 113 for January 1971 and so on. In any year January is the lowest month and December is the highest. But these are not seasonal effects, such, in fact, being entirely absent. The problem is to distinguish such cases from, for example, the monthly sales of Christmas cards, which presumably also have their lowest value in January and their highest in December, such variation being seasonal in any ordinary sense of the word.

5.2   It is as well to remind ourselves that there are different reasons for wanting to examine seasonal effects, just as there were different reasons for looking at the residual effects after the removal of trend.

(a) We may wish to compare a variable at different points of the year as a purely intra-year phenomenon; for example, in deciding how many hotels to close out of season, or at what points to allow stocks to run down.

(b) We may wish to remove seasonal effects from the series in order to study its other constituents uncontaminated by the seasonal element.

(c) We may wish to "correct" a current figure for seasonal effects, e.g. to state what the unemployment figures in a winter month would have been if customary seasonal influences had not increased them.

These objectives are not the same, and it follows that one single method of seasonal determination may not be suitable to meet them all. This is, perhaps, the reason why in practice we find different agencies (especially in Government) favouring different techniques for dealing with the seasonal problem.

Types of model

5.3   We shall consider three types of model, according as the seasonal effect is additive or multiplicative. If m_t is the smooth component of the series (trend and short-term oscillation), s_t is the seasonal component and ϵ_t the error term, we may have either

$$u_t = m_t + s_t + \epsilon_t \tag{5.1}$$

or

$$u_t = m_t s_t \epsilon_t \tag{5.2}$$

or, perhaps, a mixed model,

$$u_t = m_t s_t + \epsilon_t. \tag{5.3}$$

The purely multiplicative model (5.2) may be converted to linear form by taking logarithms

$$\log u_t = \log m_t + \log s_t + \log \epsilon_t. \tag{5.4}$$

There is a small theoretical point to mention here. If ϵ_t in (5.2) were a variable with an assignable frequency function, we could not allow its sign to be negative, except in the unlikely case where random events reverse the sign of u_t. The residual element $\log \epsilon_t$ in (5.4) will not have the same frequency function as ϵ_t itself. If we decide (as we usually do in choosing (5.4)) to regard the residual itself as a random variable η_t with zero mean

$$\log u_t = \log m_t + \log s_t + \eta_t, \tag{5.5}$$

then the original form is

$$u_t = m_t s_t e^{\eta_t}. \tag{5.6}$$

5.4   To estimate s_t in the additive model we shall first have to estimate m_t. A moving average with roughly equal weights taken over the course of the year should not affect the residuals s_t very much because, almost by definition, the sum of seasonal effects over the year is zero (or, to put it another way, any excess of the sum of seasonal effects can be absorbed into m_t). To fix the ideas, let us speak in terms of months. We shall therefore (for reasons explained in **3.11**) take a centred average

$$\tfrac{1}{24}[1, 2, 2, \ldots, 2, 2, 1] \tag{5.7}$$

and impose the condition that (for the additive model)

$$\sum_{j=1}^{12} s_t = 0. \tag{5.8}$$

The effect of this is that the simple moving average, which we should in the ordinary way expect to account for a linear trend, actually eliminates a quadratic one; and generally, an average which would eliminate a p-ic will, under the restriction (5.8), eliminate a $(p + 1)$ic. The point will be clear from an arithmetical example which, for the purpose of simplicity, we will express in quarters.

Example 5.1

Consider a set of quarterly figures and a quadratic series as follows:

(1) Year	(2) Quarter	(3) Series	(4) Centred average of 4 $\frac{1}{8}[1, 2, 2, 2, 1]$	(5) Col.(3)−col.(4)
1	1	0		
	2	1		
	3	4	5·5	− 1·5
	4	9	10·5	− 1·5
2	1	16	17·5	− 1·5
	2	25	26·5	− 1·5
	3	36	37·5	− 1·5
	4	49	50·5	− 1·5
3	1	64	65·5	− 1·5
	2	81	82·5	− 1·5
	3	100	101·5	− 1·5
	4	121	122·5	− 1·5
4	1	144	145·5	− 1·5
	2	169	170·5	− 1·5
	3	196		
	4	225		

If there were a seasonal effect added to the series in column (3), its mean under averaging would be approximately zero and it would be reproduced as an addition to column (5). We should not then take the values of column (5) as seasonal effects because the sum of four values over the year averages − 1·5 per quarter, not zero. We should then subtract − 1·5 from the values in column (5) to get an estimate of the seasonal effect. This gives the same result for seasonality as if we had added − 1·5 to the centred average in column (4), which would then reproduce the original quadratic exactly.

Example 5.2

Table 5.1 gives the quarterly index numbers of the wholesale price of vegetable food in the United Kingdom for the years 1951−8. For arithmetic convenience the scale is multiplied by 10 and the series then transferred to origin 300 in Table 5.2.

Table 5.1 *Quarterly index numbers of the wholesale price of vegetable food in the United Kingdom, 1951–8 (data from the Journal of the Royal Statistical Society for appropriate years; 1867–1877 = 100)*

	1951	1952	1953	1954	1955	1956	1957	1958
First quarter	295·0	324·7	372·9	354·0	333·7	323·2	304·3	312·5
2nd quarter	317·5	323·7	380·9	345·7	323·9	342·9	285·9	336·1
3rd quarter	314·9	322·5	353·0	319·5	312·8	300·3	292·3	295·5
4th quarter	321·4	332·9	348·9	317·6	310·2	309·8	298·7	318·4

Table 5.2 *Data of Table 5.1 with origin 300, values multiplied by 10*

	1951	1952	1953	1954	1955	1956	1957	1958
First quarter	−50	247	729	540	337	232	43	125
2nd quarter	175	237	809	457	239	429	−141	361
3rd quarter	149	225	530	195	128	3	− 77	−45
4th quarter	214	329	489	176	102	98	− 13	184

Table 5.3 gives the residuals after elimination of trend by a centred average of fours. The mean values for each quarter (the arithmetic average over seven years) are shown in the last column. These means sum to 24·01, itself with a mean of 6·00. Thus the seasonal effects are measured by subtracting 6·00 from the last column, e.g. 68·46 − 6·00. After division by 10 to restore the original scale we have for the four quarters

$$6·25, \quad 8·62, \quad −8·84, \quad −6·03, \tag{5.9}$$

which sum to zero as required.

To correct the indices for seasonality we should then, for example, subtract 6·25 from the first quarter and add 6·03 to the last.

Table 5.4 gives the similar residuals obtained by fitting a seven-point cubic $\frac{1}{21}[−2, 3, 6, 7, \ldots]$. The seasonal adjustments will be found to be

$$6·81, \quad 6·87, \quad −8·07, \quad −5·61. \tag{5.10}$$

The differences between the two results are not very great. The seasonal effect is perceptible but not very marked.

5.5   As we have expounded the subject, it has been implicit that the trend value should be determined at every point of the series except the end-group which is missed by the centred average. It is very remarkable, however, as Durbin (1963) was the first to point out, that this is unnecessary (though it may be desirable for other reasons). We develop the argument for monthly data, although it is quite general.

Table 5.3 *Residuals in the data of Table 5.2 after removal of trend by a centred moving average of fours*

	1951	1952	1953	1954	1955	1956	1957	1958	Totals	Means
First quarter		25·750	167·875	77·875	108·625	24·875	52·250	22·000	479·250	68·46
2nd quarter		−8·125	189·750	75·875	28·250	238·000	107·875	229·375	645·250	92·18
3rd quarter	−10·125	−94·750	−85·625	−121·625	−60·375	−163·875	−40·250		−576·625	−82·38
4th quarter	10·000	−122·500	−59·000	−88·000	−97·000	26·000	−49·250		−379·750	−54·25

Table 5.4 *Residuals in the data of Table 5.2 after removal of trend by a cubic fitted to 7 points*

	1951	1952	1953	1954	1955	1956	1957	1958	Totals	Means
First quarter		+30·38	122·14	79·95	113·19	16·33	91·14	5·43	458·56	65·51
2nd quarter		+29·19	135·62	82·24	27·43	206·57	−84·24		396·81	66·13
3rd quarter		−53·71	−123·95	−106·81	−35·76	−191·57	12·38		−499·42	−83·24
4th quarter	−12·67	−128·67	−72·57	−70·48	−97·90	25·81	−54·00		−410·48	−58·64

Let x_t be the deviation of u_t from a centred 12-monthly average

$$x_t = u_t - \tfrac{1}{24}\{u_{t-6} + 2u_{t-5} + \ldots + 2u_{t+5} + u_{t+6}\}. \tag{5.11}$$

Let $\bar{x}_1, \bar{x}_2, \ldots, \bar{x}_{12}$ be the monthly means of the x's, i.e.

$$\bar{x}_i = \frac{1}{p} \sum_{j=1}^{p} x_{12j+i}, \quad i = 1, 2, \ldots, 6$$

$$= \frac{1}{p} \sum_{j=1}^{p-1} x_{12j+i}, \quad i = 7, 8, \ldots, 12, \tag{5.12}$$

where the number of years is $(p + 1)$, not p.

The different limits in the summation arise from the fact that our centred average does not start to yield values until $t = 7$ and that it ends at $12p + 6$. We define the seasonal effects as

$$s_i = \bar{x}_i - \bar{x}, \tag{5.13}$$

where

$$\bar{x} = \frac{1}{12} \sum_{i=1}^{12} \bar{x}_i. \tag{5.14}$$

Let us now define monthly averages of the original series by

$$\bar{u}_i = \frac{1}{p} \sum_{j=1}^{p} u_{12j+i}, \quad i = 1, 2, \ldots, 12, \tag{5.15}$$

and the overall mean (apart from the six values at the beginning and at the end) by

$$\bar{u} = \frac{1}{12} \sum_{t=7}^{12p+6} u_t = \frac{1}{12} \sum_{j=1}^{p} \bar{u}_i. \tag{5.16}$$

Then from (5.12) we find

$$\bar{x}_1 = \frac{1}{p} \sum_{j=1}^{p} x_{12j+1}$$

$$= \frac{1}{p} \sum_{j=1}^{p} [u_{12j+1} - \tfrac{1}{24}\{u_{12j-5} + 2u_{12j-4} + \ldots + 2u_{12j+6} + u_{12j+7}\}]$$

$$= \bar{u}_1 - \frac{1}{24p} \left\{ u_7 + 2 \sum_{j=8}^{12p+6} u_t + u_{12p+7} \right\}$$

$$= \bar{u}_1 - \bar{u} + \tfrac{1}{24}(u_7 + u_{12p+7}). \tag{5.17}$$

Generally we find

$$\bar{x}_i = \bar{u}_i - \bar{u} + \frac{1}{24} \left\{ u_{i+6} + 2 \sum_{j=7}^{i+5} (u_j - u_{12p+j}) - u_{12p+i+6} \right\}, \quad i = 1, 2, \ldots, 6 \tag{5.18}$$

$$= \bar{u}_i - \bar{u} + \frac{1}{24p} \left\{ -u_{i-6} + 2 \sum_{j=i-5}^{6} (-u_j + u_{12p+j}) + u_{12p+i-6} \right\} \tag{5.19}$$

$$i = 6, 7, \ldots, 12.$$

The remarkable thing about these formulae is that $\bar{x}_i$, the deviation from the trend, can be obtained from the monthly means (about the overall mean) with no trend fitting; the only thing necessary is a correction based on the first and last thirteen terms of the series. The other $12p + 26$ values play no *individual* part in the determination of seasonal effects, except of course through the monthly means.

5.6   On the face of it this looks as if the above method of determining seasonal effects might be rather inefficient. Durbin shows that this is not so. If we consider a purely seasonal effect plus a random residual ϵ, the estimates of the monthly seasonals taken from $12p$ values without trend elimination would be 11 var $\epsilon/12p$. On the above method the variance of no seasonal constant would be affected by more than 0.7 var ϵ/p^2.

Example 5.3

Consider a set of quarterly figures for p years, $u_1, u_2, \ldots, u_{4p}$. The deviations from trend of the first quarter are

$$u_5 - \tfrac{1}{8}(u_3 + 2u_4 + 2u_5 + 2u_6 + u_7)$$
$$u_9 - \tfrac{1}{8}(u_7 + 2u_8 + 2u_9 + 2u_{10} + u_{11})$$
$$\cdot \quad \cdot \quad \cdot \quad \cdot \quad \cdot \quad \cdot \quad \cdot \quad \cdot$$
$$u_{4p-3} - \tfrac{1}{8}(u_{4p-5} + 2u_{4p-4} + 2u_{4p-3} + 2u_{4p-2} + u_{4p-1}). \qquad (5.20)$$

Their total will be found to be

$$S_1 - u_1 - \tfrac{1}{8}(2S - 2u_1 - 2u_2 - u_3 - u_{4p-1} - 2u_{4p}), \qquad (5.21)$$

where S is the sum of the whole series from 1 to $4p$ and S_1 is the sum of the values of u in the first quarters, u_1 to u_{4p-3}. Similarly, the sums in the other three quarters are

$$S_2 - u_2 - \tfrac{1}{8}\{2S - 2u_1 - 2u_2 - 2u_3 - u_4 - u_{4p}\}. \qquad (5.22)$$
$$S_3 - u_{4p-1} - \tfrac{1}{8}\{2S - u_1 - u_{4p-3} - 2u_{4p-2} - 2u_{4p-1} - 2u_{4p}\}. \qquad (5.23)$$
$$S_4 - u_{4p} - \tfrac{1}{8}\{2S - 2u_1 - u_2 - u_{4p-2} - 2u_{4p-1} - 2u_{4p}\}. \qquad (5.24)$$

The mean of these four values (since $S = S_1 + S_2 + S_3 + S_4$) is)

$$-\tfrac{1}{32}\{u_1 + 3u_2 - 3u_3 - u_4 - u_{4p-3} - 3u_{4p-2} + 3u_{4p-1} + u_{4p}\}. \qquad (5.25)$$

The seasonal factors, obtained by subtraction of this mean from the quarterly sums, are $1/p$ times

$$S_1 - \tfrac{1}{4}S - \tfrac{1}{32}\{23u_1 - 11u_2 - u_3 + u_4 + u_{4p-3} + 3u_{4p-2} - 7u_{4p-1} - 9u_{4p}\} \qquad (5.26)$$
$$S_2 - \tfrac{1}{4}S - \tfrac{1}{32}\{-9u_1 + 21u_2 - 5u_3 - 3u_4 + u_{4p-3} + 3u_{4p-2} - 3u_{4p-1} - 5u_{4p}\}$$
$$\qquad (5.27)$$

$$S_3 - \tfrac{1}{4}S - \tfrac{1}{32}\{- 5u_1 - 3u_2 + 3u_3 + u_4 - 3u_{4p-3} - 5u_{4p-2} + 21u_{4p-1} - 9u_{4p}\}$$
$$(5.28)$$

$$S_4 - \tfrac{1}{4}S - \tfrac{1}{32}\{- 9u_1 - 7u_2 + 3u_3 + u_4 + u_{4p-3} - u_{4p-2} - 11u_{4p-1} + 23u_{4p}\}.$$
$$(5.29)$$

Thus the seasonal factors are the deviations of quarterly averages from the overall average "corrected" by terms depending on only the first four and the last four terms of the series.

These corrections are nevertheless very important and must on no account be neglected. Consider, for instance, the data of Example 5.1. Arranged by year, the values are

| | | Quarter | | |
Year	1	2	3	4
1	0	1	4	9
2	16	25	36	49
3	64	81	100	121
4	144	169	196	225
Totals	224	276	336	404

Total 1240, mean 310.

The deviations $S_i - \tfrac{1}{4}S$ are $-86, -34, 26, 94$ and division by 3 would give "seasonal" factors $-28\cdot7, -11\cdot3, 8\cdot7, 31\cdot3$ if we neglected the end corrections. However, there are no seasonal effects. Applying (5.27) we have for the first quarter

$$-86 - \tfrac{1}{32}\{(23 \times 0) - (11 \times 1) - (4) + (9 \times 144) + (3 \times 169) - (7 \times 196) - (9 \times 225)\}$$
$$= -86 + 2752/32$$
$$= 0,$$

and similarly the seasonal effects for the other quarters will be found to be zero.

5.7 If we regard the seasonal element as multiplicative, some modification in procedure is required. Instead of seeking elements which are added to or subtracted from the series, we look for factors to multiply and divide and usually express them in percentage form. A seasonal factor of 110 for January would, for example, imply that we had to divide the actual January figures by $1\cdot1$ to correct for seasonality. As in the additive cases, we first determine a trend and then divide into the actual values to give an estimate of the seasonal effect.

There is one point to note here. If the results for, say, twelve months give us a set of twelve values ranging around 100, we adjust them so as to have a mean of 100. We still require the *sum* of the seasonal effects over the year to be 100, not their product. There is, perhaps, an element of illogicality about adjusting a set of quotients by adding or subtracting from each an amount to compel their mean to be 100. It seems unlikely that any method would resolve the problem completely, and it is difficult to think of a better one.

5.8   Up to this point we have considered the seasonal effects to be constant in time. If there is any reason to suppose that the seasonal pattern itself is changing, the individual seasonal factors themselves may need fitting to a trend effect. In any case it may be desirable to smooth them by a moving average.

The Bureau of the Census program

5.9   For a great many practical purposes where monthly or quarterly data are involved, use may be made of a powerful program known as Census Mark II devised by Shiskin for the U.S. Bureau of the Census (see Shiskin, 1967, and later Bureau publications). It is widely used and, notwithstanding its vulnerability on a few theoretical points, seems to work very well in practice. It purports to separate off the seasonal and residual variation, but it does *not* dissect the smooth component into trend and short-term oscillation. There are several versions which differ in minor particulars, but basically the procedure for monthly data is as follows:

(1) An option is offered whether to adjust the series for number of trading or working days. If it is adopted, all subsequent operations are on the adjusted series.

(2) A moving average is taken. There are a number of options in the extent and weighting of the average.

(3) This is divided into the series to give a first estimate of the seasonal-plus-irregular component. End values are estimated, usually as the nearest value for the same month. Extreme values are replaced by the mean of the two values for that month lying on either side of it.

(4) To decide the relative importance of seasonal and irregular components, an analysis of variance is carried out between years, between months, and residual. If the variance between months is significant, on an F-test, as compared with the residual variance, there is evidence of genuine seasonal effects.

(5) For any month the ratio of within-month to residual variance for that month is allowed to decide among a number of options what moving

average shall be used to smooth the random—residual term. A different
average may be used for different months.

(6) In some cases the seasonal so determined is divided into the primary
series to get a preliminary deseasonalised series, and another moving
average taken to get a second estimate of the trend. The seasonal factors
are adjusted so as to sum to 12.

(7) These results provide estimates of the smooth component and of a moving seasonal component. The residual is obtained by subtraction (or sometimes by dividing the primary series by smooth-plus-seasonal component
if the error is regarded as multiplicative).

(8) Various subsidiary statistics such as the error variance are computed.

Example 5.4

Fig. 5.1 shows the smooth, seasonal, and irregular factors for the airline
data of Table 1.3. The analysis on which these diagrams were based was
carried out on a Univac 1108 by a program adapted from the X–11 variant
of the Census Mark II. No adjustment was made for trading days or length
of month. A first 15-term trend-oscillation was fitted. The resulting values were
divided into the original series to get an estimate of the irregular-plus-seasonal
component. From the trend-oscillation an average from month to month,
regardless of sign, was compared with the average from month to month of
the estimated irregular-plus-seasonal component. This comparison dictated a
second moving average of the original series, also in this present case a 15-term
average, and again the trend-oscillation estimates are divided into the original
series to give the seasonal-plus-irregular component. The latter are smoothed
by a 7-term moving average, to give preliminary estimates of seasonal and
irregular component. For each month separately these preliminary estimates
are used to select a further smoothing, in this present case a 5-term average.
This provides an estimate of the seasonal components. They are divided into
the seasonal-plus-irregular component to give the irregular component, which
thus appears as multiplicative.

5.10   There are two other methods of ascertaining seasonal components,
one based on regressions and the other on harmonic analysis. For the sake
of completeness we should also mention at this point that some forecasting
methods also incorporate a seasonal component, although this is rather different from the measurement of a seasonal effect throughout the series. An
account of the method of harmonics will have to wait until we have treated
spectrum analysis in Chapter 8. We conclude this chapter with an account of
the regression method and some general comments.

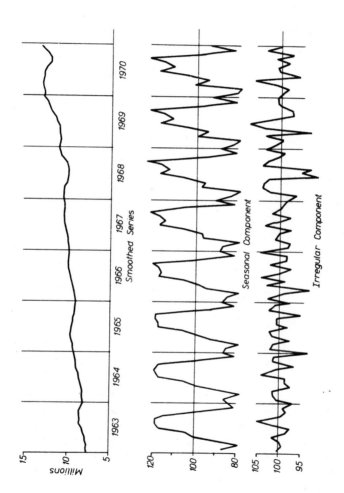

Fig. 5.1 Smooth, seasonal and irregular factors for the airline data of Table 1.3

The regression method

5.11   With a slight change of notation, let u_{ij} be the values of the original series for the ith year and the jth month. Let x_{ij} be the estimate of trend based on a centred 12-point moving average. Then we calculate the regressions

$$u_{ij} = a_j + b_j x_{ij} \tag{5.30}$$

for each of the 12 months. If $a_j = 0$ we should have a multiplicative model. If $b_j = 1$ we should have an additive model.

We then adjust the constants a and b to

$$a'_j = a_j - \bar{a}, \quad \text{where } \bar{a} = \tfrac{1}{12} \sum_{j=1}^{12} a_j \tag{5.31}$$

$$b'_j = b_j - \bar{b} + 1, \quad \text{where } \bar{b} = \tfrac{1}{12} \sum_{j=1}^{12} b_j. \tag{5.32}$$

The seasonally adjusted values are then given by

$$u_{ij} \,(\text{adjusted}) = \frac{u_{ij} - a'_j}{b'_j}. \tag{5.33}$$

This model has the advantage that it copes with both additive and multiplicative effects. However, there are serious disadvantages which would outweigh the advantage for many purposes. It requires the estimation of 24 constants, and unless the number of years is substantial those estimates are rather unreliable. The constants a' and b' are estimated for the whole series and make no allowance for rapid recent changes in seasonal pattern. And consequently the estimators are troublesome to update. The method has not come into general use.

5.12   In practical time-series analysis there are few golden rules, and indeed few general rules which can be applied without detailed thought about the nature of the series and the purpose of studying it. However, experience would indicate that it is better to get rid of known effects before proceeding to further study. I have already offered this advice in regard to "cleaning up" the series before proceeding to analyse it. For similar reasons I prefer to estimate and eliminate seasonal effects at the outset, and not to leave them in the series to get in the way, so to speak, of the subsequent analysis. This view can be justified by the consideration that seasonals are due to identifiable causes.

5.13   The same considerations do not necessarily apply to the irregular component. If we are concerned with describing the smooth component of the series, as in Fig. 5.1, it may be desirable to remove the irregular component. But if we are interested in the short-term oscillation it is probably better to remove the trend but to leave the irregular component with the short-term movement, as unaffected as possible by our methods of elimination; for we know little *a priori* about the generator mechanism of the oscillations, and it may be that the irregular components play a considerable part in it.

NOTES

(1) The Census Mark II method has been widely used outside the United States; it can deal with monthly and quarterly data but has not been adapted for weekly data. In different hands a number of variations have grown up, and the reader who wishes to use the program would do well to study the manual of instruction for the particular version he has in mind, and to insist that the print-out specifies which of the many options open to the machine have been used.

(2) The method makes no assumption that the seasonal component can be represented by harmonics, although it has to make some assumptions to dissect trend from seasonality. This may be an advantage in cases where the seasonal element is highly skew. e.g. in the build-up of demand and a sudden termination (Christmas cards, Easter eggs, certain taxation situations, etc.). In cases of doubt several methods should be tried.

6

Stationary series

6.1 We now consider a series from which trend has been removed, or in which it was never present. Such a series is called *stationary*, but there are limitations to be placed on the kind of stationarity we shall consider. In terms of intuitive ideas, we require the series to have a constant mean and to fluctuate about that mean with a constant variance. The series may then be said to be "stationary in mean and variance". More generally, if the series is u_t we may require that u_t has the same frequency distribution for any value of t, and more generally still, that any consecutive set of values $u_{t+1}, \ldots, u_{t+k}$ has the same multivariate distribution for any value of t and any value of k. This is a mathematical expression of the assumption that the generating mechanism of the series, albeit of a probabilistic kind, remains constant through time.

6.2 Under these assumptions we have

$$E(u_t) = \mu, \quad \text{say,} \tag{6.1}$$

and
$$E(u_t - \mu)^2 = \sigma^2, \quad \text{say,} \quad = \operatorname{var} u, \tag{6.2}$$

$$E(u_t - \mu)(u_{t+k} - \mu) = \gamma_k, \quad \text{the } k\text{th autocovariance,} \tag{6.3}$$

with the corresponding autocorrelation

$$\rho_k = \rho_{-k} = \gamma_k/\sigma^2. \tag{6.4}$$

The use of Expectation here is to be noticed. In the ordinary theory of probability it refers to the mean of a random variable, and so here. However, we shall have to estimate the constants, not from a set of values drawn at random from a probability distribution u, but from a set of u_t's which all have the same univariate distribution but are correlated. We shall see later that, under fairly general conditions, the average of what u_t would have been at any one point of time, had we been able to repeat the circumstances which gave rise to it, can be calculated from the range of values actually observed over a period of time.

6.3   On the same point, we may remark that if we are regarding the series as a sample of all the series of that length which might have been generated by the same mechanism, we have, in a sense, only a sample of one. It is conveniently described as a *realization* of the underlying mechanism, which can be called a *stochastic process*. What we have just been saying is that we can estimate the parameters of a stochastic process from one realization of it. We shall, in fact, estimate the means, variances and covariances of a process from the corresponding statistics computed from the realization. The conditions under which this is legitimate are contained in a theorem of Birkhoff (1931) and Khintchine (1932). We recall that a process has a range of autocorrelations given by (6.4) for all values of k. Then, if m is the limit of the mean of a set of n terms as n tends to infinity, (we state without proof) m exists for almost all realizations (namely its existence has probability unity); and if further

$$\lim_{n \to \infty} \frac{1}{n} \sum_{j=1}^{n} \rho_j = 0, \tag{6.5}$$

then $m = \mu$. The most common cases are those in which either ρ_k vanishes after some point k, or ρ_k dwindles to zero rapidly enough for the average of ρ_k to tend to zero.

The correlogram

6.4   The suite of values ρ_k and the diagrammatic representation of it are both known as the *correlogram*. All correlations, of course, must not lie outside the limits ± 1, but not every sequence of constants within those limits can be a correlogram. Consider, in fact, three values u_t, u_{t+1}, u_{t+2}. If we seek for the partial correlation of u_t and u_{t+2} when their dependence on u_{t+1} is removed, we have, in virtue of a well-known formula,

$$\rho_{13.2} = \frac{\rho_{13} - \rho_{12} \rho_{23}}{(1 - \rho_{12}^2)^{\frac{1}{2}}(1 - \rho_{23}^2)^{\frac{1}{2}}}$$

$$= \frac{\rho_2 - \rho_1^2}{1 - \rho_1^2}.$$

If this is to lie within limits ± 1 we find

$$-1 + 2\rho_1^2 \leqslant \rho_2 \leqslant 1. \tag{6.6}$$

The right-hand inequality is trivial, but the left-hand one is not. If, for example, $\rho_1 = 0 \cdot 8$, ρ_2 must not be less than $0 \cdot 28$, a reflection of the fact that if successive values in the series are highly positively correlated, those terms one unit apart have not had time, as it were, to become uncorrelated or negatively correlated.

Autoregression

6.5   One important class of stationary series which we have already met in another context is that generated on a random series by a moving average of finite extent. In Table 3.2 we gave an expression for the autocorrelations in terms of the weights of the moving average, and in Fig. 3.1 we graphed the correlogram of a series generated by the Spencer 21-point formula. We now consider another important class known, for obvious reasons, as *auto-regressive*.

The autoregressive scheme of order k with constant coefficients is defined by the equation

$$u_{t+k} + \alpha_1 u_{t+k-1} + \alpha_2 u_{t+k-2} + \ldots + \alpha_k u_t = \epsilon_{t+k} \tag{6.7}$$

or equivalently

$$u_t = -\alpha_1 u_{t-1} - \alpha_2 u_{t-2} - \ldots - \alpha_k u_{t-k} + \epsilon_t. \tag{6.8}$$

In the form (6.8) we can regard u_t as regressed on u_{t-1}, u_{t-2}, with a random residual. (It does not follow, as we shall see later, that standard regression theory applies to such series.) The series may be considered as generated by a mechanism in which the value of the series at time t is expressed in terms of the past values — a systematic dependence on past history — plus a "disturbance" term ϵ happening at time t. We consider two important simple cases, known by the names of Markoff and Yule.

The Markoff scheme

6.6 The Markoff scheme, the simplest linear autoregressive scheme other than the purely random series, is defined by

$$u_t = -\alpha_1 u_{t-1} + \epsilon_t, \tag{6.9}$$

which we may more conveniently write as

$$u_t = \rho u_{t-1} + \epsilon_t. \tag{6.10}$$

Using the same formula with $t-1$ instead of t, we find

$$\begin{aligned} u_t &= \epsilon_t + \rho \epsilon_{t-1} + \rho^2 u_{t-2} \\ &= \epsilon_t + \rho \epsilon_{t-1} + \rho^2 \epsilon_{t-2} + \rho^3 u_{t-3}, \end{aligned} \tag{6.11}$$

and so on. Evidently u_t depends on ϵ_t and previous ϵ_t's but not on future ϵ's. If ϵ has zero mean, so has u_t. Then, on multiplying (6.10) by u_{t-1} and taking expectations, we have

$$\operatorname{cov}(u_t, u_{t-1}) = \rho \operatorname{var} u, \tag{6.12}$$

and so the first autocorrelation of the system is ρ, which justifies our introduction of that symbol.

Similarly, on multiplying (6.10) by u_{t-k} and taking expectations, we have

$$\operatorname{cov}(u_t, u_{t-k}) = \rho \operatorname{cov}(u_{t-1}, u_{t-k})$$

and on division by var u

$$\rho_k = \rho\rho_{k-1} = \rho^2\rho_{k-2}, \text{etc.,}$$
$$= \rho^k. \tag{6.13}$$

All the autocorrelations of the Markoff scheme are therefore expressible in terms of its first autocorrelation ρ. The diagrammatic form of the correlogram for $\rho > 0$ is given later (Fig. 6.3).

6.7   A Markoff series, in appearance, presents oscillations of a more or less regular kind. By the use of the result of **3.18** we can find the mean distance between its peaks. The cosine of the angle between

$$u_{t+1} - u_t = 0 \quad \text{and} \quad u_t - u_{t-1} = 0$$

is

$$\frac{E\{(u_{t+1} - u_t)(u_t - u_{t-1})\}}{\{E(u_{t+1} - u_t)^2 E(u_t - u_{t-1})^2\}^{\frac{1}{2}}} = \frac{2\rho - 1 - \rho^2}{2(1-\rho)} = -\tfrac{1}{2}(1 - \rho). \tag{6.14}$$

We then have that the mean distance between peaks is

$$2\pi/\{\text{arc cos} \left[-\tfrac{1}{2}(1 - \rho)\right]\}. \tag{6.15}$$

For example, with $\rho = 0$ (the random series),$\cos^{-1} -\tfrac{1}{2} = \tfrac{2}{3}\pi$, so the mean distance is 1·5, confirming what we found in Chapter 2. For $\rho = \tfrac{1}{2}$ the mean distance is $2\pi/\cos^{-1}(-\tfrac{1}{4}) = 360/104·5 = 3·45$.

6.8   We can continue equations like (6.11) as far back in time as we like. Since $|\rho|$ is of necessity less than unity, the coefficients diminish in importance, and if we stop at the term u_{t-k} its coefficient is ρ^k. For many purposes it is more convenient to consider the series as stretching back into the infinite past:

$$u_t = \sum_{j=0}^{\infty} \rho^j \epsilon_{t-j}. \tag{6.16}$$

If the series was, in fact, started up at some unknown past point, say k units ago, the error we commit in expressing u_t in this way is negligible for moderate ρ and k. Formula (6.16) exhibits u_t as a moving average of infinite extent, but with weights summing, not to unity, but to $(1 - \rho)^{-1}$.

6.9   The reader who is familiar with the Central Limit Theorem of statistics, which states that under general conditions the mean of a weighted sum of variables with any distribution tends to normality, might be inclined to infer from (6.16) that u_t, being such a sum, is normally distributed. This is not so. In fact, one of the conditions for the validity of the Central Limit Theorem is violated: the weights in the average are not of the same order of magnitude.

From (6.10) we see that u_t is the sum of variables ρu_{t-1} and ϵ_t which are independent. The logarithm of the characteristic function of u_t is therefore the sum of the logarithms of the characteristic functions of ρu_{t-1} and ϵ_t. Equivalently, if κ_r is the rth cumulant,

$$\kappa_r(u_r) = \rho^r \kappa_r(u_{t-1}) + \kappa_r(\epsilon),$$

and since $\kappa(u_t) = \kappa(u_{t-1})$ this gives us

$$\kappa_r(u_t) = \frac{\kappa_r(\epsilon)}{1 - \rho^r}. \tag{6.17}$$

In particular,

$$\text{var } u = \frac{\text{var } \epsilon}{1 - \rho^2}, \tag{6.18}$$

a formula which is easily verified directly.

6.10 One important consequence of (6.18) is that if $|\rho|$ is near to unity, var u will be much larger than var ϵ. If the successive values of the series are highly correlated, a series of quite small disturbances will generate wide swings.

A second consequence is that if we standardize the cumulants of (6.18) by dividing κ_r by $\kappa_2^{r/2}$, we find

$$\frac{\kappa_r(u)}{\kappa^{\frac{1}{2}r}(u)} = \frac{\kappa_r(\epsilon)}{\kappa^{\frac{1}{2}r}(\epsilon)} \frac{(1 - \rho^2)^{\frac{1}{2}r}}{1 - \rho^r}. \tag{6.19}$$

In general the factor $(1 - \rho^2)^{\frac{1}{2}r}/(1 - \rho^r)$ is less than unity and tends to zero as r tends to infinity. Thus the standardized cumulants of u are smaller than those of ϵ and its distribution is accordingly closer to normality. If ϵ is symmetrically distributed so is u, for odd order cumulants then vanish.

The Yule scheme

6.11 The Yule autoregressive process is defined by

$$u_t = -\alpha_1 u_{t-1} - \alpha_2 u_{t-2} + \epsilon_t. \tag{6.20}$$

As in the Markoff case, ϵ_t is independent of u_{t-1} and u_{t-2}. Multiplying (6.20) by these in turn and taking expectations, we have

$$\left.\begin{array}{l} \rho_1 + \alpha_1 + \alpha_2 \rho_1 = 0 \\ \rho_2 + \alpha_1 \rho_1 + \alpha_2 = 0. \end{array}\right\} \tag{6.21}$$

These equations are enough to determine α_1 and α_2 in terms of the first two autocorrelations. We find

$$\rho_1 = -\frac{\alpha_1}{1 + \alpha_2} \tag{6.22}$$

$$\rho_2 = -\alpha_2 + \frac{\alpha_1^2}{1 + \alpha_2}, \tag{6.23}$$

and conversely

$$\alpha_1 = -\frac{\rho_1(1 - \rho_2)}{1 - \rho_1^2} \tag{6.24}$$

$$\alpha_2 = -1 + \frac{1 - \rho_2}{1 - \rho_1^2} = -\frac{\rho_2 - \rho_1^2}{1 - \rho_1^2}. \tag{6.25}$$

Generally, multiplying (6.20) by u_{t-k}, $k \geqslant 1$, we find a range of equations

$$\rho_k + \alpha\rho_{k-1} + \alpha_2\rho_{k-2} = 0. \tag{6.26}$$

These can, if desired, be used to determine higher-order ρ's in terms of the first two, which therefore determine the correlogram completely.

6.12   We can also regard (6.26) as a difference equation in ρ of the second order with constant coefficients. Very much as for a differential equation of the second kind, it may be shown that the general solution is

$$\rho_j = A\mu^j + B\nu^j \tag{6.27}$$

where μ, ν are the roots of

$$\cdot \quad x^2 + \alpha_1 x + \alpha_2 = 0 \tag{6.28}$$

and the initial conditions are, for $j = 0, 1, -1$ (since, as is seen from (6.21), (6.26) also holds for $j = -1$),

$$1 = A + B$$

$$\rho_1 = A\mu + B\nu = \rho_{-1} = \frac{A}{\mu} + \frac{B}{\nu}.$$

We then find

$$\rho_j = \frac{1}{(\mu - \nu)(1 + \mu\nu)} \{\mu^{j+1}(1 - \nu^2) - \nu^{j+1}(1 - \mu^2)\}. \tag{6.29}$$

When the roots of (6.28) are complex, we can put this into a rather more convenient form. Put

$$\mu = pe^{i\theta}, \quad \nu = pe^{-i\theta}.$$

Then we find

$$p = |\sqrt{\alpha_2}|, \quad \cos\theta = -\frac{\alpha_1}{2|\alpha_2|}, \tag{6.30}$$

and (6.29) reduces to

$$\rho_j = \frac{p^j \sin(j\theta + \psi)}{\sin\psi}, \tag{6.31}$$

where

$$\tan\psi = \frac{1 + p^2}{1 - p^2} \tan\theta. \tag{6.32}$$

The form of (6.31) shows that the correlogram is a damped harmonic. A typical graph is given later (Fig. 6.4).

6.13   The mean distance between peaks in a Yule series can be ascertained by the method we have already employed on the random and Markoff schemes. If $v_t = u_{t+1} - u_t$,

$$\text{var } v_t = 2(1 - \rho_1) \text{ var } u.$$
$$\text{cov } (v_{t+1}, v_t) = (-1 + 2\rho_1 - \rho_2) \text{ var } u.$$

Thus the correlation between v_{t+1} and v_t is given by

$$\text{corr } (v_{t+1}, v_t) = \frac{-1 + 2\rho_1 - \rho_2}{2(1 - \rho_1)}. \tag{6.33}$$

If this is $\cos \theta$ the mean distance between peaks is $2\pi/\theta$. On substituting from (6.22) to (6.23), we find the simple form

$$\text{corr } (v_{t+1}, v_t) = \tfrac{1}{2}(\alpha_2 - \alpha_1 - 1). \tag{6.34}$$

If $\alpha_2 = 0$ this reduces to the result we have already found for the Markoff scheme. For example, with $\alpha_1 = -1 \cdot 1, \alpha_2 = 0 \cdot 5$, we find $\theta = \cos^{-1} 0 \cdot 3 = 72 \cdot 5°$ The mean distance between peaks is $360/72 \cdot 5$, about 5.

6.14   In Fig. 6.1 we have graphed a series of 65 terms of the Yule scheme with $\alpha_1 = -1 \cdot 1, \alpha_2 = 0 \cdot 5$, the random element ϵ being a rectangular variable with range from $-9 \cdot 5$ to $+9 \cdot 5$. The first point to define a peak is the 4th and the last is the 59th. In two places adjacent values have the same peak point. We therefore count $65 - 3 - 5 - 2 = 55$ points to define a peak. There are 12 in the series, giving a mean distance of $4 \cdot 6$, in good agreement with theory, although the random element is not normal.

For comparison, Fig. 6.2 gives the values of 60 terms of a harmonic series generated by

$$u_t = 10 \sin \frac{\pi t}{5} + \epsilon_t, \tag{6.35}$$

where ϵ is a rectangular random variable ranging from -5 to $+5$. The period of this series is 10. Comparison with Fig. 6.1 shows the different appearances of the two kinds of series, the harmonic having more or less regular peaks and approximately constant amplitude, the Yule scheme varying in both.

6.15   Incidentally, it may be remarked that a "peak" as we have defined it is only a value greater than the two adjacent values, and may be a very minor concern, as witness some of the peaks in Fig. 6.1. The counting of peaks is not a very sensitive diagnostic as between a random, a Markoff and a Yule process. In the random series we expect an interval of about 3, in the Markoff with positive ρ a value between 3 and 4, and in the Yule scheme a value somewhere between 4 and 6.

6.16   It was obvious that in the Markoff case the parameter ρ cannot exceed unity; otherwise the variance would increase without limit and the series

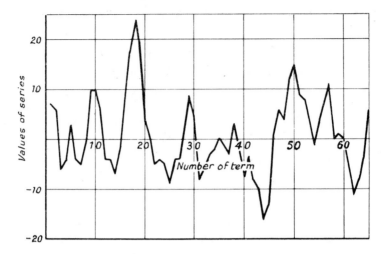

Fig. 6.1   Graph of 65 terms of a second-order autoregressive Yule scheme

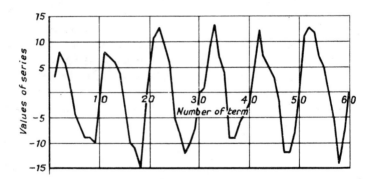

Fig. 6.2   Graph of 60 terms of a harmonic series

could "explode". Similarly there are limitations on the values of α_1 and α_2 for the stationarity of a Yule scheme. They are that the roots of

$$x^2 + \alpha_1 x + \alpha_2 = 0$$

shall both have modulus less than unity. For if not, the autocorrelations as given in (6.27) would sooner or later exceed unity in absolute value. In fact, the necessary and sufficient conditions for stationarity are

$$|\alpha_1| < 2 \quad \text{and} \quad -\alpha_2 < 1 - |\alpha_1|. \tag{6.36}$$

6.17   The general linear autoregressive scheme can be written

$$\sum_{j=1}^{h} \alpha_j u_{t-j} = \epsilon_t. \tag{6.37}$$

If B is a backward shift operator defined by $Bu_t = u_{t-1}$, we may write this as

$$(\Sigma \, \alpha_j B^j)u_t = \epsilon_t. \tag{6.38}$$

The solution of this equation, regarded as a difference equation in u_t, consists of two parts — a general solution, containing h arbitrary constants, plus a particular solution. The general solution is

$$u_t = \sum_{j=1}^{h} A_j \, z_j^t, \tag{6.39}$$

where the z_j are the h roots of the equation

$$\Sigma \, \alpha_j \, z^{h-j} = 0. \tag{6.40}$$

For the series to remain stationary, the z's (although possibly complex) must have modulus less than unity, which is sometimes expressed by saying that they must lie inside the unit circle. In general we shall suppose that the series was started up some time ago, so that (6.39) has damped out of existence. The remaining part of the solution is

$$u_t = \frac{1}{\Sigma \, \alpha_j B^j} \epsilon_t = (\Sigma \, \beta_j B^j) \, \epsilon_t$$

$$= \sum_{j=0}^{\infty} \beta_j \, \epsilon_{t-j}, \tag{6.41}$$

where the α's are connected with the β's by the identity in B

$$1 = (\Sigma \, \alpha_j B^j)(\Sigma \, \beta_j B^j), \tag{6.42}$$

from which the β's can be calculated from the α's if required.

6.18   Equation (6.41) reveals the linear autoregressive scheme as a moving average of infinite extent. The series ϵ_t will be stationary and we state without proof that it will obey the criterion of (6.5) if $\Sigma \, \beta_j^2$ converges, which will be so if the roots of (6.39) lie within the unit circle, as may be shown by expressing $1/\Sigma \, \alpha_j B^j$ in partial fractions and expanding.

6.19   Multiplying equation (6.37) by u_{t-k}, $k > 0$ and taking expectations, we have a suite of equations

$$\rho_k + \alpha_1 \rho_{k-1} + \ldots + \alpha_h \rho_{k-h} = 0, \quad k > 0. \tag{6.43}$$

These are known as the Yule–Walker equations; we have already used them for the Yule and Markoff processes. It follows from them that the autocorrelations of the series are all determined by the first h. We shall find considerable use of them in sampling problems.

6.20   A particular case occurs when one or more roots of (6.39) lie on the unit circle. The series is then not stationary, but "wanders" with increasing variance. For example, the "Markoff" with

$$u_t = u_{t-1} + \epsilon_t$$

has clearly

$$u_t = \sum_{j=0}^{\infty} \epsilon_{t-j}$$

and its variance is not contained within bounds.

Partial autocorrelation

6.21   We may extend the idea of autocorrelation, which measures the correlation of terms of the series separated by assigned numbers of terms, to that of the correlation where dependence on the intermediate terms has been removed. Consider, for example, a Markoff scheme. The term u_{t+2} is correlated with u_t to extent ρ^2. But u_{t+2} depends on u_{t+1} and the latter on u_t. Is u_{t+2} correlated with u_t solely in virtue of the fact that both are correlated with u_{t+1}?

Denoting the terms in order by 1, 2, 3, we have, in an obvious notation, for the partial correlation

$$\rho_{13.2} = \frac{\rho_{13} - \rho_{12}\rho_{32}}{(1 - \rho_{12}^2)^{\frac{1}{2}}(1 - \rho_{32}^2)^{\frac{1}{2}}}. \tag{6.44}$$

Now $\rho_{13} = \rho^2$, and ρ_{12} and ρ_{23} are both equal to ρ, so the numerator vanishes. Likewise a partial correlation such as $\rho_{14.23}$ will depend in its numerator on $\rho_{14.2}$ and $\rho_{14.3}$, which are both seen to vanish. In fact all the partials vanish. Thus the autocorrelations are solely due to the correlation of neighbouring terms. This is otherwise evident from the fact that u_t is defined in terms of u_{t-1} alone.

6.22   Consider now the Yule scheme. Using the formula (6.44) with values of the ρ's in terms of α_1 and α_2, from (6.24) and (6.25) we find that

$$\rho_{13.2} = \frac{\rho_2 - \rho_1^2}{1 - \rho_1^2} = -\alpha_2. \tag{6.45}$$

Higher-order partials vanish. For example, the numerator $\rho_{14.23}$ in expressions like (6.44) is $\rho_{14.3} - \rho_{12.3}\rho_{42.3}$. This in turn has numerator

$$(1 - \rho_1^2)(\rho_3 - \rho_2\rho_1) - (\rho_1 - \rho_2\rho_1)(\rho_2 - \rho_1^2). \tag{6.46}$$

Considering the determinant of the first three Yule–Walker equations of (6.43), we have

$$\begin{vmatrix} \rho_1 & 1 & \rho_1 \\ \rho_2 & \rho_1 & 1 \\ \rho_3 & \rho_2 & \rho_1 \end{vmatrix} = 0$$

and expansion by the first column shows that (6.46) vanishes.

6.23   The general result may be seen in the same way. For a scheme of order h the partial correlations of terms h or more apart vanish. In fact, the correlation matrix of a set of p u's, $p \geqslant h$, is

$$
\begin{bmatrix}
1 & \rho_1 & \rho_2 & \rho_3 & . & \rho_p \\
\rho_1 & 1 & \rho_1 & \rho_2 & . & \rho_{p-1} \\
\rho_2 & \rho_1 & 1 & \rho_1 & . & \rho_{p-2} \\
. & . & . & . & . & . . \\
\rho_p & \rho_{p-1} & \rho_{p-2} & \rho_{p-3} & . & . \; 1
\end{bmatrix}
\tag{6.47}
$$

The partial correlation of variables $1, h + 1$ with the intermediate values fixed is given by a general formula (Kendall and Stuart, vol. 2, chapter 27) in which the numerator is the minor of the term in the first row and $(h + 1)$t column. This, expanded in terms of the $h \times h$ determinants on its left, must vanish, for each determinant vanishes in virtue of the Yule–Walker equations.

6.24   We can, of course, graph the partial autocorrelations to form a partial correlogram. Whereas the correlogram of an autoregressive series may extend to infinity, the partial correlogram of the right order should vanish, just as does the correlogram of a finite moving-average process, from some point onwards. Unfortunately, as we shall see, this does not give us a very reliable diagnostic in practice owing to sampling effects in serial correlations. In fact, the explicit calculation of partials by formulae of type (6.44) can be avoided for autoregressive schemes. Cf. **12.7**.

6.25   In addition to the moving-average process (a term which we confine to averages of finite extent) and the autoregressive process, we shall also have to consider mixed schemes; for example

$$
\sum_{j=0}^{h} \alpha_j\, u_{t-j} \; = \; \sum_{0}^{k} \beta_j\, \epsilon_{t-j}.
\tag{6.48}
$$

We can look on this as an autoregressive scheme in which the residual errors are themselves correlated, the correlations being generated by a moving average of independent terms. More complicated models can be constructed but are difficult to handle, and wherever possible we must aim at simplicity in the form chosen to represent a stationary process.

6.26   For experimental purposes it is often desirable to construct autoregressive schemes from random elements such as those given in tables of random numbers or random deviates. The question then arises, how should we begin, since the first value u_1 depends on u_0 and possibly previous values which are unknown. There are two ways of proceeding.

(1) We may assume u_0 and previous terms to be zero and start the series up "from cold". It should then be run for a number of terms until the effect of these initial assumptions has dwindled to negligible importance. The series up to that point should be discarded as a transient unrepresentative segment.

(2) We may compute the ratio var u/var ϵ and start with a random value ϵ but multiply it by (var u/var $\epsilon)^{\frac{1}{2}}$ so as to give it the right variance for u. This series "warms up" quicker than that obtained by the first method, but it is still desirable to discard a few of the initially calculated values of u.

For the Markoff scheme we have already found that

$$\frac{\text{var } u}{\text{var } \epsilon} = \frac{1}{1 - \rho^2}.$$

For the Yule scheme we have

$$\text{var } u = \text{E}(u_t + \alpha_1 u_{t-1} + \alpha_2 u_{t-2})^2$$
$$= \text{var } u\{1 + \alpha_1^2 + \alpha_2^2 + 2\alpha_1\rho_1 + 2\alpha_2\rho_2 + 2\alpha_1\alpha_2\rho_1\}$$

and on substituting for ρ_1, ρ_2 in terms of α_1 and α_2,

$$\frac{\text{var } u}{\text{var } \epsilon} = \frac{1 + \alpha_2}{(1 - \alpha_2)\{(1 + \alpha_2)^2 - \alpha_1^2\}}. \qquad (6.49)$$

6.27   It may be worth remarking that, although the autocorrelations uniquely determine the constants in an autoregressive scheme through the medium of the Yule–Walker equations, this is not true of a moving-average scheme. A very simple example is the scheme $\frac{1}{4}[1, 3]$, which gives the same autocorrelation of first order as $\frac{1}{4}[3, 1]$, further correlations vanishing. Another example, due to Wold (1954), is given by the four schemes $\frac{1}{5}[8, -4, 2, -1]$, $\frac{1}{5}[-1, 2, -4, 8], \frac{1}{5}[2, -1, 8, -4], \frac{1}{5}[-4, 8, -1, 2]$, all of which have the first three autocorrelations as $-42/85, 4/17, -8/85$ and the rest zero.

The spectrum

6.28   In statistical theory the characteristic function, defined in terms of the frequency function $f(x)$ by

$$\phi(t) = \int_{-\infty}^{\infty} f(x)\, e^{itx} dx, \qquad (6.50)$$

can profitably be used as a moment-generating function. On expansion formally of exp (itx) in terms of t we have

$$\phi(t) = \sum_{r=0}^{\infty} \mu_r' \frac{(it)^r}{r!},$$

where μ'_r is the rth moment about the origin. We now define a function which, among other properties, can be regarded as a generator of the autocovariances. We define the *spectral density* $w(\alpha)$ with argument α (not to be confused with the coefficients α in an autoregressive series as in **6.17**) by

$$w(\alpha) = \sum_{j=-\infty}^{\infty} \rho_j e^{ij\alpha} \qquad (6.51)$$

which, in virtue of the fact that $\rho_j = \rho_{-j}$, may be written as the real expression

$$w(\alpha) = 1 + 2 \sum_{j=1}^{\infty} \rho_j \cos \alpha j. \qquad (6.52)$$

If we multiply this by $\cos \alpha k$ and integrate term by term with respect to α over the range 0 to π, we find, since

$$\int_0^\pi \cos \alpha j \cos \alpha k \, d\alpha = 0, \quad j \neq k$$
$$= \tfrac{1}{2}\pi, \quad j = k,$$
$$\int_0^\pi w(\alpha) \cos \alpha k \, d\alpha = \pi \rho_k, \qquad (6.53)$$

so that
$$\rho_k = \frac{1}{\pi} \int_0^\pi \cos k\alpha \, w(\alpha) d\alpha. \qquad (6.54)$$

The correlogram and the spectral density uniquely determine each other.

We may also define the *spectral function* as an integrated form $W(\alpha)$ of $w(\alpha)$:

$$W(\alpha) = \int_0^\alpha w(\alpha) d\alpha = \alpha + 2 \sum_{j=1}^{\infty} \rho_j \frac{\sin \alpha j}{j}, \qquad (6.55)$$

the series being convergent. To exhibit the symmetry of the relation between the correlogram and the spectrum we may also write

$$\rho_k = \frac{1}{\pi} \int_0^\pi w(\alpha) \, e^{-ik\alpha} \, d\alpha. \qquad (6.56)$$

6.29   The graph of $w(\alpha)$ as ordinate against α as abscissa is called the *spectrum*. $w(\alpha)$ has period 2π, but since $\cos(2\pi - \alpha)j$ is the same as $\cos \alpha j$ the spectrum is symmetrical about $\alpha = \pi$, so only the graph from 0 to π is necessary.

From (6.55) we see that $W(0) = 0$, $W(\pi) = \pi$, $W(2\pi) = 2\pi$.

It is sometimes convenient to graph the spectrum with $\log w(\alpha)$ instead of $w(\alpha)$ as ordinate.

6.30   We shall study the spectrum from a different viewpoint in Chapter 8, but we may notice here that it arises naturally in a manner in which autocorrelations do not appear.

Consider, in fact, the relationship between the observed series measured about its mean and a harmonic term of period $2\pi/\alpha$. Let

$$a(\alpha) = \frac{1}{\sqrt{(n\pi)}} \sum_{t=1}^{n} u_t \cos \alpha t, \tag{6.57}$$

$$b(\alpha) = \frac{1}{\sqrt{(n\pi)}} \sum_{t=1}^{n} u_t \sin \alpha t. \tag{6.58}$$

We have

$$I(\alpha) = a^2(\alpha) + b^2(\alpha) = \frac{1}{n\pi} \{(\Sigma u_t \cos \alpha t)^2 + (\Sigma u_t \sin \alpha t)^2\}$$

$$= \frac{1}{n\pi} \left[\Sigma u_t^2 + 2 \sum_{k=1}^{n-1} \sum_{t=1}^{n-k} \{\cos \alpha t \cos \alpha(t+k) \right.$$
$$\left. + \sin \alpha t \sin \alpha(t+k)\} u_t u_{t+k} \right]$$

$$= \frac{1}{n\pi} \left\{ \Sigma u_t^2 + 2 \sum_{k=1}^{n-1} \sum_{t=1}^{n-k} u_t u_{t+k} \cos k\alpha \right\}$$

$$= \frac{S^2}{\pi} \left\{ 1 + 2 \sum_{k=1}^{n-1} r_k \cos k\alpha \right\},$$

where S^2 is the observed variance $\Sigma u_t^2/n$ and r_k is a type of serial correlation $\Sigma u_t u_{t+k}/\Sigma u_t^2$. We say a "type" because there are $n - k$ terms summed in the numerator and n in the denominator; to make it a sample serial correlation we should multiply by $n/(n-k)$, but in the limit, to which we now proceed, that factor is unity. In the limit, then,

$$E\left\{ I(\alpha) \right\} = \frac{\sigma^2}{\pi} \left\{ 1 + 2 \sum_{k=1}^{\infty} \rho_k \cos k\alpha \right\} \tag{6.59}$$

which, from (6.52), is seen to be equivalent to

$$E\left\{ I(\alpha) \right\} = \frac{\sigma^2}{\pi} w(\alpha). \tag{6.60}$$

The spectral density $w(\alpha)$, apart from the constant σ^2/π, measures the strength of the relationship between u and the harmonic of period $2\pi/\alpha$. $I(\alpha)$ is called the *intensity*.

Autocorrelation generating functions

6.31   In the spectral density function

$$w(\alpha) = \sum_{-\infty}^{\infty} \rho_j e^{i\alpha j}$$

put

$$z = e^{i\alpha}. \tag{6.61}$$

Then

$$w(\alpha) = \sum_{-\infty}^{\infty} \rho_j z^j. \tag{6.62}$$

Thus $w(\alpha)$, expanded as a power series in z, gives the autocorrelations. Conversely, given the autocorrelations we can write down the spectrum.

Example 6.1   Spectrum of the Markoff series

From equation (6.62) we have at once for the Markoff series

$$\Sigma \, \rho_j z^j \; = \; \overset{\infty}{\underset{-\infty}{\Sigma}} \, \rho^{|j|} z^j \; = \; \frac{1}{1 - \rho z} + \frac{1}{1 - \rho z^{-1}} - 1,$$

and replacing z by $e^{i\alpha}$,

$$w(\alpha) \; = \; \frac{1}{1 - \rho e^{i\alpha}} + \frac{1}{1 - \rho e^{-i\alpha}} - 1$$

$$= \; \frac{1 - \rho^2}{1 - 2\rho \, \cos \alpha + \rho^2} \, . \tag{6.63}$$

Fig. 6.3 shows the correlogram and spectrum of a typical Markoff series for positive ρ.

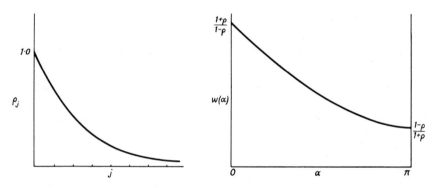

Fig. 6.3   Correlogram (*left*) and spectrum (*right*) of the Markoff infinite series

6.32   Consider two series, one of which is a moving sum of the other:

$$y_t \; = \; \overset{\infty}{\underset{i=0}{\Sigma}} \, \alpha_i \, x_{t-i}. \tag{6.64}$$

We have written this as an infinite sum, but the finite sum is included as a particular case when the α's vanish from some point onwards. Then

$$\mathrm{E}\,(y_t y_{t+j}) \; = \; \mathrm{E} \left\{ \overset{\infty}{\underset{i=0}{\Sigma}} \, \alpha_i x_{t-i} \right\} \left\{ \overset{\infty}{\underset{i=0}{\Sigma}} \, \alpha_i x_{t+j-i} \right\}$$

$$= \; \overset{\infty}{\underset{i=0}{\Sigma}} \, \overset{\infty}{\underset{k=0}{\Sigma}} \, \alpha_i \alpha_k \, \mathrm{E}\, x_{t-i} x_{t+j-k}$$

$$= \; \Sigma \, \Sigma \, \alpha_i \alpha_k \, \rho_{|\,i+j-k\,|} \, \mathrm{var} \; x$$

$$= \; \overset{\infty}{\underset{i=0}{\Sigma}} \, \overset{\infty}{\underset{l=-\infty}{\Sigma}} \, \alpha_i \alpha_{i+j-l} \rho_l \, \mathrm{var} \; x,$$

where ρ_l is the autocorrelation of x. It will then be seen that the coefficient of the jth autocovariance of y is that of z^j in

$$(1 + \alpha_1 z + \alpha_2 z^2 + \ldots)(1 + \alpha_1 z^{-1} + \alpha_2 z^{-2} + \ldots)(\text{autocovariance generator of } x).$$

Thus, if $A(z)$ is the autocovariance generator of x and $B(z)$ is that of y,

$$B(z) = \left(\sum_0^\infty \alpha_j z^j\right)\left(\sum_0^\infty \alpha_j z^{-j}\right) A(z). \qquad (6.65)$$

This is an important result which we shall frequently use.

Example 6.2   Spectrum of the Yule scheme

Let y be a random variable ϵ, and x_t our usual u_t. Then, since the autocovariance generator of a random series reduces to var ϵ, we have for the Yule scheme

$$\text{var } \epsilon = (1 + \alpha_1 z + \alpha_2 z^2)(1 + \alpha_1 z^{-1} + \alpha_2 z^{-2})A(z) \text{ var } u, \qquad (6.66)$$

so that the spectral density is given by

$$w(\alpha) = \frac{\text{var } \epsilon}{\text{var } u} / \{1 + \alpha_1^2 + \alpha_2^2 + \alpha_1(1 + \alpha_2)(e^{i\alpha} + e^{-i\alpha}) + \alpha_2(e^{2i\alpha} + e^{-2i\alpha})\}$$

$$= \frac{\text{var } \epsilon}{\text{var } u} / \{1 + \alpha_1^2 + \alpha_2^2 + 2\alpha_1(1 + \alpha_2) \cos \alpha + 2\alpha_2 \cos 2\alpha\}. \qquad (6.67)$$

We found the ratio var ϵ/var u at equation (6.49). Fig. (6.4) shows the theoretical correlogram and the spectrum of the Yule (i.e. second-order autoregressive) series.

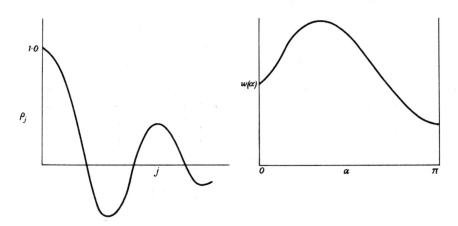

Fig. 6.4   Correlogram (*left*) and spectrum (*right*) of the Yule infinite series

6.33 It will follow in the same way that the autocovariance generator of the mixed autoregressive—moving average scheme of (6.48) is given by

$$\frac{(\Sigma \beta_j z^j)(\Sigma \beta_j z^{-j})}{(\Sigma \alpha_j z^j)(\Sigma \alpha_j z^{-j})} \text{ var } \epsilon \tag{6.68}$$

and its spectral density by putting, as usual, $z = e^{i\alpha}$.

6.34 The theory of time-series, especially in connection with spectrum analysis, has imported two terms used in engineering. A typical situation confronting a communications engineer is one in which he has an incoming signal (a stationary series) and transforms it by some device to an outgoing signal, usually to remove noise or to emphasize certain elements in the signal so as to transmit a pattern of information. This transformation is known as a *filter*. A purist, perhaps, might object that it is not a filter because it does not, as a rule, hold back any material but only hands it on in a modified form. However, the term is convenient and well-established. If the transformation is a linear one it becomes, in effect, a moving average. The engineer, however, can also consider continuous processes anu write down integral expressions in place of what we have written as finite sums. Thus, for example, if we have a weight function $a(t)$ and a function $u(t)$ we may form the moving average

$$v(t) = \int_0^\infty a(\tau)u(t-\tau)d\tau, \tag{6.69}$$

the continuous variable τ taking the place of our summation over discontinuous (and usually integral values) of t.

The transfer function

6.35 We may define a transform for such a function, again in integral form, as

$$\left.\begin{aligned} \int_{-\infty}^\infty v(t)e^{it\alpha}\,dt &= \int_0^\infty a(\tau)\int_{-\infty}^\infty e^{it\alpha}u(t-\tau)dt\,d\tau \\ &= \int_0^\infty a(\tau)e^{i\tau\alpha}\int_{-\infty}^\infty u(t)e^{it\alpha}\,dt\,d\tau. \end{aligned}\right\} \tag{6.70}$$

The function

$$T(\alpha) = \int_0^\infty a(\tau)e^{i\tau\alpha}d\tau \tag{6.71}$$

(which can also be defined over a doubly infinite range) is called the *transfer function*. It transforms the Fourier transform of the original series to that of the outgoing series. Since it is amplitudes rather than phase which are of interest, the expression is usually stated in terms of the modulus, so that if $w_v(\alpha)$ refers to v,

$$|w_v(\alpha)| = |T^2(\alpha)|\,|w_u(\alpha)|. \tag{6.72}$$

The objects of the engineer, in designing good transfer functions to confer satisfactory properties on his outgoing signal, are analogous to those of the time-series analyst, except perhaps that the latter is as much interested in the elements which are filtered off as in those that remain.

NOTES

(1) The ordinate of the spectrum as defined will be seen from (6.60) to be non-negative. It may, however, be zero. If the logarithm of $w(\alpha)$ is taken as ordinate, the range may therefore extend to $-\infty$. Small values of $w(\alpha)$ are not usually of interest, and in graphical presentation the lower part of the ordinate scale is sometimes omitted.

(2) The constants of an autoregressive scheme of order h can, as remarked in the text, be derived from the first h Yule–Walker equations. There is, however, an infinite set of equations and, for a finite realization of n terms, there are a great many more than h. The question arises whether we should use them in estimating the constants of the series. The answer is in the negative (cf. 12.4). The point was discussed experimentally in Kendall (1949).

(3) When the autoregressive parameters α are regarded as expressing some internal relationship of the system under study, it is natural to consider the case when they themselves vary with time. Not very much study has been devoted to models with time-dependent parameters; if, for example, the α's are considered as polynomials in t, the series ceases to be stationary. A simpler way of treating such cases is to work with the differences of the series in a model with constant coefficients (cf. 9.21), though this is not necessarily nearer to the true structure of the system.

7

Problems in sampling serial correlation and correlogram

7.1   The sampling theory of individual serial correlations, and *a fortiori* of the whole correlogram, is in general rather complicated from the mathematical point of view and involves some considerations which are not present in most distributional theory. A fairly complete account is given in Kendall and Stuart, vol. 3. In the present chapter we shall concentrate on presenting the basic ideas and results of practical importance, being content to quote a number of the more advanced results without proof. We begin with some large-sample theory, which itself is sufficient for many purposes.

Definitions

7.2   We noted in Chapter 3 that there are various ways in which a serial correlation can be defined, one exact and the others a fairly close approximation. For large n they tend to equivalence, and we are at liberty to choose the simplest. Moreover we can suppose the series measured about its parent mean, not the sample mean, without loss of accuracy in determining sampling variances to order $1/n$. Thus we define

$$r_j = \frac{\Sigma \, u_i u_{i+j}}{\Sigma \, u_i^2} = \frac{c}{v}. \tag{7.1}$$

In accordance with the usual procedure for determining the standard error of a function of variables in terms of their individual standard errors we have, writing δ for a small deviation,

$$\delta r_j = \frac{\delta c}{v} - \frac{c \delta v}{v^2}. \tag{7.2}$$

Squaring and taking expectations gives us

$$\text{var } r_j = \frac{\text{var } c}{v^2} - \frac{2 \, \text{cov } (v, c)}{v^3} + \frac{c^2 \, \text{var } v}{v^4}. \tag{7.3}$$

Without loss of generality we may now take $v = 1$. We thus require the variance of c and v and their covariance. Actually we shall obtain a more general expression for the covariance of two terms of covariance type. We have, to our present order of magnitude,

$$E\left\{\frac{1}{n}\Sigma u_a u_{a+s}\right\} = \rho_s.$$

Hence

$$\text{cov}\left\{\frac{1}{n}\Sigma u_a u_{a+s} \frac{1}{n}\Sigma u_b u_{b+s+t}\right\}$$

$$= \frac{1}{n^2}E\left\{\Sigma u_a u_{a+s} u_b u_{b+s+t}\right\} - \rho_s \rho_t$$

$$= \frac{1}{n^2}\sum_{a,b}E(u_a u_{a+s} u_b u_{b+s+t}) - \rho_s \rho_t. \qquad (7.4)$$

This expression requires the expectation of the product of four u's and therefore depends on one of the fourth-order moment. To avoid this we now impose the condition that the u's are normally distributed. Their characteristic function is then of the form

$$\exp\left\{\tfrac{1}{2}(\theta_a^2 + \theta_{a+s}^2 + \theta_b^2 + \theta_{b+s+t}^2 + 2\rho_s \theta_a \theta_{a+s} + \text{similar products})\right\}, \qquad (7.5)$$

where θ is the imaginary parameter usually written as it.

The coefficient of $\theta_a \theta_{a+s} \theta_b \theta_{b+s+t}$ is the fourth-order moment required, and this is easily found to be

$$\rho_s \rho_{s+t} + \rho_{b-a}\rho_{b-a+t} + \rho_{b-a+s+t}\rho_{b-a-s}. \qquad (7.6)$$

Insertion in (7.4) and summation gives us for the covariance

$$\frac{1}{n}\left\{\Sigma \rho_i \rho_{i+t} + \Sigma \rho_{i+s+t}\rho_{i-s}\right\}. \qquad (7.7)$$

Specializing to $s = t = 0$, we have

$$\text{var } v = \frac{2}{n}\sum_{-\infty}^{\infty} \rho_i^2. \qquad (7.8)$$

Putting $t = 0$,

$$\text{var } c = \frac{1}{n}\sum_{-\infty}^{\infty} (\rho_i^2 + \rho_{i+s}\rho_{i-s}). \qquad (7.9)$$

Putting $s = 0$ and replacing t by s,

$$\text{cov }(v, c) = \frac{2}{n}\sum_{-\infty}^{\infty} \rho_i \rho_{i+s}. \qquad (7.10)$$

Finally, from (7.3),

$$\text{var } r_j = \frac{1}{n}\sum_{-\infty}^{\infty} [\rho_i^2 + \rho_{i-j}\rho_{i+j} - 4\rho_i \rho_j \rho_{i+j} + 2\rho_i^2 \rho_j^2]. \qquad (7.11)$$

7.3   This formula (due to Bartlett, 1946) exhibits one of the difficulties which face us. The variance of r depends on *all* the autocorrelations of the series, however long. The consequence is that we have to specialize by assuming something about the generative mechanism of the system. (We have already assumed that the variation is normal.) For a moving-average scheme, for example, all correlations above a certain order vanish. For an autoregressive scheme of order h they can all be expressed in terms of the first h correlations.

Example 7.1

For a random series with all correlations equal to zero we find

$$\text{var } r_j = \frac{1}{n}. \tag{7.12}$$

For a scheme in which ρ_j and subsequent ρ's are small, an approximation to (7.11) is

$$\text{var } r_j = \frac{1}{n} \sum_{-\infty}^{\infty} \rho_i^2. \tag{7.13}$$

For the Markoff scheme, for large j,

$$\text{var } r_j = \frac{1}{n} \sum_{-\infty}^{\infty} \rho^{2i}$$
$$= \frac{1}{n}\left\{\frac{1+\rho^2}{1-\rho^2}\right\}. \tag{7.14}$$

7.4 It is noteworthy that successive values of r_j are correlated, and may be highly so. Analogously to (7.13) for the Markoff scheme,

$$\text{cov }(r_j, r_{j+k}) = \frac{1}{n} \sum_i \rho_i \rho_{i+k}$$
$$= \frac{1}{n}\rho^k\left\{k-1+\frac{2}{1-\rho^2}\right\}. \tag{7.15}$$

The correlation between r_j and r_{j+k} is then approximately

$$\frac{\rho^k\{(k+1)-(k-1)\rho^2\}}{1+\rho^2}. \tag{7.16}$$

For example, with $\rho = \frac{1}{2}$, r_j and r_{j+1} are correlated, with coefficient 0·80.

7.5 The correlations between neighbouring values in an observed correlogram, together with the size of the standard error, account for one regrettable feature of empirical correlograms, namely that they do not damp out as required by theory. Fig. 7.1 will illustrate the point. It is the correlogram of the Yule series of 65 terms referred to above and graphed in Fig. 6.1. For an infinite series the correlogram should damp out as in Fig. 6.4. Actually it preserves the oscillatory effect but fails to damp; and this is typical of series of 100 or less.

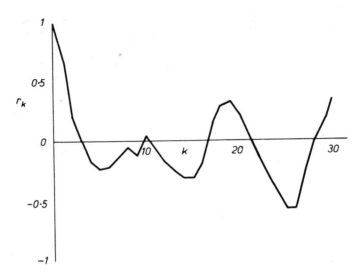

Fig. 7.1   Correlogram of the artificial Yule scheme of 65 terms (Fig. 6.1)

7.6   More exact tests of individual correlations for normally distributed u can be derived by considering expectations or by direct sampling theory. However, there is a further price to pay, namely that we usually have to assume that there are no autocorrelations in the parent, i.e. that it is a random normal series. This, unfortunately, is the case of least interest, which is why we shall not enter into the subject deeply in this book. Some close approximations are also available for the Markoff case, but results for Yule and higher-order processes are not.

(a) In normal samples and zero autocorrelation we have for the first serial correlation defined by

$$r_1 = \frac{n}{n-1} \frac{\sum\limits_{i=1}^{n-1} (u_i - \bar{u})(u_{i+1} - \bar{u}))}{\sum\limits_{i=1}^{n} (u_i - \bar{u})^2}$$

$$E(r_1) = -\frac{1}{n-1} \tag{7.17}$$

$$\text{var } r_1 = \frac{(n-2)^2}{(n-1)^3} = \frac{1}{n+1} + O(n^{-2}). \tag{7.18}$$

(Moran, 1947/8)

In fact, for any parent with zero autocorrelations,

$$\text{var } r_1 \leqslant \frac{1}{n-1} + O(n^{-2}). \tag{7.19}$$

(Moran, 1967)

(b) By defining the serial coefficient to be circular (i.e. taking u_{n+1} to be u_1), as

$$r_{1c} = \frac{u_1 u_2 + u_2 u_3 + \ldots + u_n u_1 - n\bar{u}^2}{\sum_{i=1}^{n} (u_i - \bar{u})^2}, \tag{7.20}$$

R.L. Anderson was able (1942) to obtain an explicit form for the distribution of r_1 in the case of normal variation, zero autocorrelations. The distribution is cumbrous but tends to normality quite quickly:

$$E(r_{1c}) = -\frac{1}{n-1} \tag{7.21}$$

$$\text{var } r_{1c} = \frac{n(n-3)}{(n+1)(n-1)^2}. \tag{7.22}$$

About 30 years later it was discovered that some of Anderson's work had been anticipated in a most remarkable way by Ernst Abbe in 1854 — see Kendall (1971).

(c) Dixon (1944) showed that a very close approximation to the Anderson distribution is given by the Beta distribution

$$\frac{\Gamma(\frac{1}{2}n + \frac{1}{2})}{\Gamma(\frac{1}{2}n)\,\Gamma(\frac{1}{2})} (1 - r^2)^{\frac{1}{2}(n-2)} \tag{7.23}$$

for which

$$E(r_{1c}) = -\frac{1}{n-1}$$

$$\text{var } r_{1c} = \frac{n(n-2)}{(n+1)(n-1)^2}. \tag{7.24}$$

(d) Madow (1945) and Leipnik (1947) extended the results to the Markoff scheme, obtaining a distribution of type (7.23) with

$$E(r_{1c} \mid \rho) = \frac{n\rho}{n+2} \tag{7.25}$$

$$\text{var}(r_{1c} \mid \rho) = \frac{1 - \rho^2}{n}. \tag{7.26}$$

(e) Daniels (1956) carried the theory a stage further by the use of the method of steepest descents, but the formulae are formidable.

Bias in the estimation of autocorrelations

7.7   It is natural enough to estimate the parent ρ's from the corresponding sample r's, and since the variances are of order $1/n$ the estimators are consistent, provided any bias tends to zero with n. There are, however, some unexpected biases in the estimators which we proceed to discuss.

If r is of the form $A/\sqrt{(BC)}$ we have, writing a, b, c for deviations from the means,

$$r_j = \frac{E(A) + a}{\{E(B) + b\}^{\frac{1}{2}}\{E(C) + c\}^{\frac{1}{2}}}, \tag{7.27}$$

and expanding the denominator in binomial series to the second order of approximation we find

$$E(r_j) = \frac{E(A)}{\{E(B)E(C)\}^{\frac{1}{2}}} \left\{ 1 - \frac{E(ab)}{2E(A)E(B)} - \frac{E(ac)}{2E(A)E(B)} + \frac{E(bc)}{4E(B)E(C)} \right. $$
$$\left. + \frac{3E(b^2)}{8E^2(B)} + \frac{3E(c^2)}{8E^2(C)} \right\}. \tag{7.28}$$

Now specialize to

$$B = \frac{1}{n-j} \sum_{i=1}^{n-j} u_i^2 - \frac{1}{(n-j)^2} \left(\sum_{i=1}^{n-j} u_i \right)^2 \tag{7.29}$$

$$C = \frac{1}{n-j} \sum_{i=1}^{n-j} u_{i+j}^2 - \frac{1}{(n-j)^2} \left(\sum_{i=1}^{n-j} u_{i+j} \right)^2. \tag{7.30}$$

B and C are then asymptotically equivalent and $E(B) = E(C)$. Likewise $\text{cov}(bc)$ $= \text{var } b = \text{var } c$. Then (7.28) reduces to

$$E(r_j) \doteqdot \frac{E(A)}{E(B)} - \frac{\text{cov}(a, b)}{E^2(B)} + \frac{E(A) \text{ var } b}{E^3(B)}. \tag{7.31}$$

Without loss of generality, let the variance of the series be unity and write $\nu = n - j$. Then we have from (7.29)

$$E(B) = E(C) = \frac{1}{\nu} \left\{ \nu - 1 - \frac{2}{\nu} \sum_{i=1}^{\nu-1} (\nu - i)\rho_i \right\}. \tag{7.32}$$

Similarly, taking A to be

$$A = \frac{1}{n-j} \sum_{i=1}^{n-j} u_i u_{i+j} - \frac{1}{(n-j)^2} \sum_{i=1}^{n-j} u_i \sum_{i=1}^{n-j} u_{i+j}, \tag{7.33}$$

we find

$$E(A) = \frac{1}{\nu} \left\{ \nu\rho_j - \frac{1}{\nu} \sum_{i=0}^{\nu-1} (\nu - j)\rho_{j+i} - \frac{1}{\nu} \sum_{i=0}^{j} (\nu - i)\rho_{j-i} \right.$$
$$\left. - \frac{1}{\nu} \sum_{i=1}^{\nu-j-1} (\nu - j - i)\rho_i \right\}, \quad j < 0. \tag{7.34}$$

We have evaluated var b and cov (a, b) in (7.8) and (7.10). We can then evaluate $E(r_j)$ from (7.31).

Example 7.2

If the series is random, all autocorrelations vanish and we find

$$E(r_j) = -\frac{1}{\nu} = -\frac{1}{n-1}. \tag{7.35}$$

This agrees with the exact result noted in **7.6(a)**. It is remarkable that even in samples from a random series the bias is downwards. I have sometimes observed in calculated correlograms for relatively short series that the correlogram diminishes to small values but fluctuates a little below the horizontal axis. This could be due to bias in estimation.

For the Markoff scheme we find, after a little algebra,

$$E(r_1) = \rho - \frac{1+3\rho}{n-1} \tag{7.36}$$

$$E(r_j) = \rho^j - \frac{1}{n-j}\left\{\frac{1+\rho}{1-\rho\cdot}(1-\rho^j) + 2j\rho^j\right\}, \quad j > 1. \tag{7.37}$$

The bias in all these cases is downwards and often far from negligible. For instance, in a series with $\rho = \frac{1}{2}$ the mean value of r_1 in a series of 26 values would be 0·4, not 0·5.

7.8 For short series, therefore, it seems preferable to remove the bias of order $1/n$ by a device due to Quenouille (1956). If we split the series into two halves, so that r is the serial value for the whole series and $r_{(1)}, r_{(2)}$ those for the halves, then

$$R = 2r - \tfrac{1}{2}\{r_{(1)} + r_{(2)}\} \tag{7.38}$$

will be unbiased to order n^{-1}. For if

$$E(r) = \rho + \frac{k}{n} + O(n^{-2}),$$

where k is a constant independent of n,

$$E\{r_{(1)}\} = E\{r_{(2)}\} = \rho + \frac{2k}{n} + O(n^{-2}),$$

and on substitution in (7.38) we find that

$$E(R) = \rho + O(n^{-2}). \tag{7.39}$$

7.9 In actual practice we are usually more concerned with the fitting of an autoregressive, moving-average, or mixed process, namely with the correlogram as a whole, rather than with one particular ordinate, especially as the neighbouring ordinates are correlated. We use the correlograms, as well as other diagnostics such as the spectrum, to suggest what kind of model should be set up; there then remain the problems of estimating the parameters in that model and seeing whether it satisfactorily fits the data. We postpone a discussion of such points until after we have given an account of the spectrum (see Chapter 12).

NOTES

(1) Sargan (1953) gives a theoretical reason why the correlogram for relatively short series fails to damp but preserves the expected oscillatory effect.

(2) In view of the sampling variability of serial correlations it is sometimes difficult to know how many serials to calculate for a relatively short series of, say, 50 terms. It is rarely that one needs more than $\frac{1}{2}n$ or 30, whichever is the less. Cross-product terms of this order are, however, required for certain methods of computing the spectrum—cf. **8.23**. Kendall (1946) shows the effect of increasing the length of the series in stages from 60 to 480 terms.

8

Spectrum analysis

8.1 Jean-Baptiste Joseph Fourier is remembered among mathematicians for his methods of representing a function as a sum of harmonic terms. The typical Fourier series may be written

$$f(x) = \sum_{j=1}^{\infty} a_j \sin jx + \tfrac{1}{2}b_0 + \sum_{j=1}^{\infty} b_j \cos jx, \qquad (8.1)$$

or equivalently

$$f(x) = \sum_{j=0}^{\infty} c_j \sin (jx + \phi_j). \qquad (8.2)$$

The harmonic terms in which the series is expressed are of a special kind in regard to period: the first sine term has period 2π, or more conveniently $-\pi$ to π, the second has period $-\tfrac{1}{2}\pi$ to $\tfrac{1}{2}\pi$, the third $-\tfrac{1}{3}\pi$ to $\tfrac{1}{3}\pi$ and so on. Similarly for the cosine terms. The whole sum has period 2π.

Notwithstanding this periodicity, many functions can be represented by a Fourier series over the limited range $-\pi$ to π (or any finite range by a change of scale). It is sufficient for this to be possible that the function be single-valued, continuous except perhaps at a finite number of points, and possess only a finite number of maxima or minima. For example,

$$\tfrac{1}{2}x = \sin x - \tfrac{1}{2}\sin 2x + \tfrac{1}{3}\sin 3x - \tfrac{1}{4}\sin 4x + \dots \qquad 0 \leqslant x \leqslant \pi. \quad (8.3)$$

8.2 Owing to the fact that the periods of successive terms are submultiples of a basic period, we can easily determine the constants a and b. In fact,

$$\int_{-\pi}^{\pi} \cos rx \sin sx \, dx = 0$$

$$\left. \begin{aligned} \int_{-\pi}^{\pi} \cos rx \cos sx \, dx &= \int_{-\pi}^{\pi} \sin rx \sin sx \, dx \\ &= 0, \quad r \neq s \\ &= \pi, \quad r = s. \end{aligned} \right\} \qquad (8.4)$$

Multiplying $f(x)$ of (8.1) by $\sin jx$ or $\cos jx$ and integrating, we then find

$$a_j = \frac{1}{\pi} \int_{-\pi}^{\pi} f(x) \sin jx \, dx \qquad (8.5)$$

95

$$b_j = \frac{1}{\pi} \int_{-\pi}^{\pi} f(x) \cos jx \, dx. \tag{8.6}$$

Frequency and wavelength

8.3 Angles have zero dimensions when measured in radians and consequently, in an expression such as $\sin \alpha t$, the quantity α is in radians per unit of time. It is sometimes called the *angular frequency*. As $\sin \alpha t$ repeats itself with period $2\pi/\alpha$ and therefore the number of cycles per unit time is $\alpha/2\pi$, this is also sometimes called the *frequency*. Where necessary one distinguishes by speaking of angular or cycle frequency. The period $2\pi/\alpha$ is of dimension t and is called the *wavelength*. It then appears that for a function expanded in a Fourier series the successive terms have periods 2π, $2\pi/2$, $2\pi/3$, etc.; the angular frequencies are $1, 2, 3, \ldots$, and the cycle frequencies are $1/2\pi$, $2/2\pi$, $3/3\pi, \ldots$, etc. More generally, for a function defined over a time-length $2L$, the angular frequencies are typified by $j\pi/L$.

8.4 The time-series such as we encounter in real life may or may not be capable of representation as a sum of harmonic terms. We can, of course, always fit a Fourier series, but there is, in general, no reason to suppose that it corresponds to any reality — it could, for example, be entirely misleading for prediction. If, however, we do suspect the existence of harmonic components we cannot as a rule assume that their frequencies are multiples of one fundamental as in the Fourier case. We are then led to consider the more general type of series

$$f(x) = \sum_{j=0}^{\infty} \{a_j \sin(\alpha_j x) + b_j \cos(\alpha_j x)\}, \tag{8.7}$$

where the α's are not necessarily commensurable. The property expressed by equations (8.5) and (8.6) is now lost, and we cannot determine the coefficients a and b in any simple way.

8.5 In older work with representations of the type of equation (8.7) a good deal of effort was spent in trying to estimate a and b, and hence to set up a harmonic model. We shall shortly see how the spectrum can be used for that purpose. However, it is now recognized that the effects in the spectrum which would once upon a time have been held to identify harmonic terms are capable of other interpretations, and the spectrum is to be considered as an entity in itself, like the correlogram. We have already seen that theoretically either the spectrum or the correlogram can be deduced one from the other, and it might be questioned whether it is necessary to compute them both. The answer is decidedly affirmative. Except perhaps with vast experience, it is impossible to tell from the correlogram what the spectrum would look like, or vice versa, and each form presents its own difficulties in interpretation.

8.6 For series observed at equal time-intervals there are two important features to observe. Periodicities of less than one time-unit may escape notice. In fact, we remarked earlier that one way of avoiding seasonal variation was to observe the series once a year. We need at least two observations in the time-interval to show up periodicities of one unit. Generally, for a time-interval t_0 between observations we cannot detect periods smaller than $2t_0$ or angular frequencies higher than π/t_0. This limiting value is known as the *Nyquist frequency*. In the spectral density as we defined it,

$$w(\alpha) = \sum_{-\infty}^{\infty} \rho_j \, e^{i\alpha j}, \tag{8.8}$$

the time-interval is unity and the range of α is taken from 0 to π. The ordinate at π is the value of the spectral density at the Nyquist frequency.

8.7 The second effect to remark is also related to the interval of observation. Consider the term $\sin 2\pi t/3$ for unit intervals $t = 1, 2, 3$, etc. Its values are $\frac{1}{2}\sqrt{3}, -\frac{1}{2}\sqrt{3}, 0$, etc. But these are the values which would be taken by $\sin 8\pi t/3$, $\sin 14\pi t/3$, etc. The width of the interval of observation does not allow us to distinguish among the family of angular frequencies $2\pi/3 + 2\pi j$, $j = 1, 2, 3, \dots$, etc. These higher frequencies are known as *aliases*. They do not cause any trouble in most practical cases.

Intensity

8.8 In **6.30** we considered the Intensity, whose mean, but for a constant, is the spectral ordinate, as the sum of squares of two numbers a, b, defined as

$$a(\alpha) = \frac{1}{\sqrt{(n\pi)}} \sum_{t=1}^{n} u_t \cos \alpha t, \tag{8.9}$$

$$b(\alpha) = \frac{1}{\sqrt{(n\pi)}} \sum_{t=1}^{n} u_t \sin \alpha t. \tag{8.10}$$

We can look on these functions as the covariance of the series with $\cos \alpha t$ or $\sin \alpha t$ as the case may be, and we expect that if u_t contains a term with angular frequency α the intensity will be relatively large. Let us verify this. Suppose that

$$u_t = c \sin \alpha t + g_t, \tag{8.11}$$

where g_t is some stationary process not correlated with $\sin \alpha t$. We calculate a and b at one step by evaluating

$$\sum_{t=1}^{n} \sin \alpha t \, e^{i\beta t}$$

and find that it is equal to

$$\frac{1}{\sin\{\tfrac{1}{2}(\alpha-\beta)\}} \times$$

$$[\cos\{(n+\tfrac{1}{2})(\alpha-\beta)\}-\cos\{\tfrac{1}{2}(\alpha-\beta)\}-i\sin\{(n+\tfrac{1}{2})(\alpha-\beta)\}+i\sin\{\tfrac{1}{2}(\alpha-\beta)\}]$$

$$+ \text{ a similar term with } -\beta \text{ in place of } \beta. \tag{8.12}$$

In the neighbourhood of $\beta=\alpha$ this is dominated by the first term. The sum $\Sigma\, g_t e^{i\beta t}$, by hypothesis, is negligible. Hence the intensity is the sum of squares of real and imaginary parts in (8.12), multiplied by $c^2/n\pi$. It reduces to

$$I(\beta) = \frac{c^2 \sin^2\{\tfrac{1}{2}n(\alpha-\beta)\}}{4\pi n \sin^2\{\tfrac{1}{2}(\alpha-\beta)\}}. \tag{8.13}$$

Consider this in the neighbourhood where, say,

$$\alpha-\beta = \frac{2\pi m}{n}, \quad n \text{ large, } m \text{ finite.}$$

Since for small angles $\sin\theta=\theta$, it becomes

$$I(\beta) = \frac{nc^2 \sin^2 \pi m}{4\pi m^2}. \tag{8.14}$$

Thus at $\alpha=\beta$ the intensity is of the order of n, and the spectrum will have a high peak.

8.9 Used as a device for hunting down periodicities in the data, the spectrum might be compared to the tuner on a radio set. As we go along the spectrum through the range of frequencies, the intensity should remain low until we arrive at the frequency of a harmonic component of the series, at which point it shows a high peak. The collection of peaks should determine the collection of harmonic components. Unfortunately, as we show in a moment, the interpretation is far from being so straightforward.

8.10 In older literature a slightly different form of the intensity was used, based on

$$A = \frac{2}{n} \sum_{t=1}^{n} u_t \cos\frac{2\pi t}{\lambda}, \quad \lambda = 2\pi/\alpha, \tag{8.15}$$

$$B = \frac{2}{n} \sum_{t=1}^{n} u_t \sin\frac{2\pi t}{\lambda}, \tag{8.16}$$

$$S^2(\lambda) = A^2 + B^2 = \frac{4\pi I(\alpha)}{n}. \tag{8.17}$$

The differences from the ordinary spectrum are twofold:

(1) The divisors in A and B are n, not $\sqrt{(n\pi)}$. Hence the intensity at a harmonic, instead of (8.14), is given by

$$S^2 = (c^2 \sin^2 m\pi)/m^2. \tag{8.18}$$

The ordinate S^2 then remains finite and in fact equals the amplitude c^2 of the supposed harmonic, which was one reason why it was defined in that way.

(2) The values of S^2 were graphed against λ, the wavelength, as abscissa, not the frequency, so that the range extended to infinity, or at least half the length of the series.

In this form the diagram of S^2 against λ was called a *periodogram*. Occasionally one finds this last word used in the literature to denote the intensity.

Graphically, either the spectral density or the intensity are in use – the difference is only one of scale.

Example 8.1

We gave the theoretical form of the spectrum for the Markoff and Yule schemes in Chapter 6 (Fig. 6.3 and 6.4). The theoretical spectrum of a random series is simply a horizontal line with unit ordinate. The random series with small intervals between observations is sometimes known as "white noise", "noise" because it is unsystematic, "white" because it has a spectrum like that of white light.

It is of considerable interest to consider what happens to the spectrum when there is a linear trend in the data. We can replace rather tedious trigo-nometrical summations by integrals, at least for relatively small time-intervals of observation. Suppose that $u_t = t$. Then

$$\int_0^T t \sin \alpha t \, dt = -\frac{T \cos \alpha T}{\alpha} + \frac{\sin \alpha T}{\alpha^2} \tag{8.19}$$

$$\int_0^T t \cos \alpha t \, dt = \frac{T \sin \alpha T}{\alpha} + \frac{\cos \alpha T - 1}{\alpha^2}. \tag{8.20}$$

Hence

$$I(\alpha) = \frac{1}{\pi T} \left\{ \frac{T^2}{\alpha^2} + O(T) \right\} = \frac{T}{\pi \alpha^2} + O(1). \tag{8.21}$$

The spectrum with T constant will then be a curve of type $y = 1/x^2$ with a high peak at the origin. The analysis, in fact, treats the trend as if it were a long wave with zero frequency. An exact summation would give $T/4\pi \sin^2 \frac{1}{2}\alpha + O(1)$, which is equivalent for small α to (8.21).

Examples of power spectra

8.11 It is time to look at some practical examples. Fig. 8.1 presents the power spectrum of the Yule scheme graphed in Fig. 6.1 – the correlogram was given in Fig. 7.1. (The dotted line in Fig. 8.1 will be explained later.)

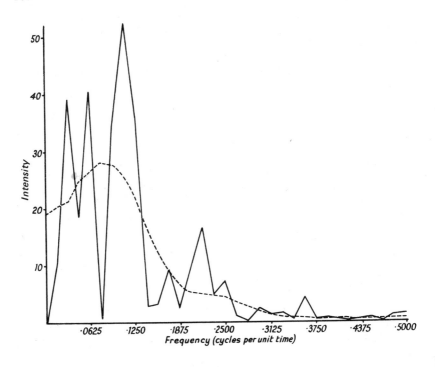

Fig. 8.1   Intensity of the Yule scheme graphed in Fig. 6.1

Table 8.1 gives one of the most famous series of our subject, the Beveridge wheat-price index for the phenomenally long period of 1500-1869; the spectrum is graphed in Fig. 8.2. In Table 8.2 we give the residuals of the immigration data of Table 1.4 after the elimination of trend by a simple moving average of 17 of the logarithm of the series. The spectrum is given in Fig. 8.3.

Observational material often presents these highly serrated spectra, and we shall see shortly why this is so.

Side-bands

8.12   Reverting to equation (8.14), we see that in the neighbourhood of an underlying harmonic of angular frequency β there will not only be a peak at β itself but a range of smaller peaks where $\sin \pi m$ attains its maxima, namely at $m = \frac{1}{2}, \frac{3}{2}$, and so on. This is obviously a nuisance for interpretation. In all three of the diagrams just presented the main peaks are flanked by smaller ones, and it is almost impossible to say whether they are side-band effects or not.

Echo effects

8.13   If there is a latent harmonic of frequency α in the series, giving a

Table 8.1 *Trend-free wheat-price index (European prices) compiled by Lord (then Sir William) Beveridge for the years 1500–1869*

Year	Index	Year	Index	Year	Index	Year	Index	Year	Index	Year	Index	Year	Index
1500	106	1553	90	1606	81	1659	104	1712	115	1765	101	1818	94
01	118	54	100	07	98	60	120	13	134	66	106	19	86
02	124	55	123	08	115	61	167	14	108	67	113	20	84
03	94	56	156	09	94	62	126	15	90	68	108	21	76
04	82	57	71	10	93	63	108	16	89	69	108	22	77
05	88	58	71	11	100	64	91	17	89	70	131	23	71
06	87	59	81	12	99	65	85	18	94	71	136	24	71
07	88	60	84	13	100	66	73	19	107	72	119	25	69
08	88	61	97	14	94	67	74	20	89	73	106	26	82
09	68	62	105	15	88	68	80	21	79	74	105	27	93
10	98	63	90	16	92	69	74	22	91	75	88	28	114
11	115	64	78	17	100	70	78	23	94	76	84	29	103
12	135	65	112	18	82	71	83	24	110	77	94	30	110
13	104	66	100	19	73	72	84	25	111	78	87	31	105
14	96	67	86	20	81	73	106	26	103	79	79	32	82
15	110	68	77	21	99	74	134	27	94	80	87	33	80
16	107	69	80	22	124	75	122	28	101	81	88	34	78
17	97	70	93	23	106	76	102	29	90	82	94	35	82
18	75	71	112	24	106	77	107	30	96	83	94	36	88
19	86	72	131	25	121	78	115	31	80	84	92	37	102
20	111	73	158	26	105	79	113	32	76	85	85	38	117
21	125	74	113	27	84	80	104	33	84	86	84	39	107
22	78	75	89	28	97	81	92	34	91	87	93	40	95
23	86	76	87	29	109	82	84	35	94	88	108	41	101
24	102	77	87	30	148	83	86	36	101	89	108	42	92
25	71	78	79	31	114	84	101	37	93	90	86	43	88
26	81	79	90	32	108	85	74	38	91	91	78	44	92
27	129	80	90	33	97	86	75	39	122	92	87	45	115
28	130	81	87	34	92	87	66	40	159	93	85	46	139
29	129	82	83	35	97	88	62	41	110	94	103	47	90
30	125	83	85	36	98	89	76	42	90	95	130	48	80
31	139	84	76	37	105	90	79	43	81	96	95	49	74
32	97	85	110	38	97	91	97	44	84	97	84	50	78
33	90	86	161	39	93	92	134	45	102	98	87	51	86
34	76	87	97	40	99	93	169	46	102	99	120	52	105
35	102	88	84	41	99	94	111	47	100	1800	139	53	138
36	100	89	106	42	107	95	109	48	109	01	117	54	141
37	73	90	111	43	106	96	111	49	104	02	105	55	138
38	86	91	97	44	96	97	128	50	90	03	94	56	107
39	74	92	108	45	82	98	163	51	99	04	125	57	82
40	74	93	100	46	88	99	137	52	95	05	114	58	81
41	76	94	119	47	116	1700	99	53	90	06	98	59	97
42	80	95	131	48	122	01	85	54	80	07	93	60	116
43	96	96	143	49	134	02	72	55	85	08	94	61	107
44	112	97	138	50	119	03	88	56	117	09	94	62	92
45	144	98	112	51	136	04	77	57	112	10	104	63	79
46	80	99	99	52	102	05	66	58	95	11	140	64	81
47	54	1600	97	53	72	06	64	59	91	12	121	65	94
48	69	01	80	54	63	07	69	60	88	13	96	66	119
49	100	02	90	55	76	08	125	61	100	14	96	67	118
50	103	03	90	56	75	09	175	62	97	15	130	68	93
51	129	04	80	57	77	10	108	63	88	16	178	69	102
52	100	05	77	58	103	11	103	64	95	17	126		

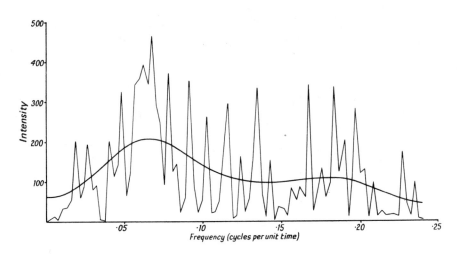

Fig. 8.2   Intensity of the Beveridge wheat-price index series (Table 8.1)

spectral peak at α, there are also, of course, periodic terms of frequency $\frac{1}{2}\alpha$, $\frac{1}{3}\alpha$, etc. A monthly periodicity may then be expected to show peaking effects with two-monthly, three-monthly, etc., periods — it "echoes", so to speak, down the spectrum. Fig. 8.4 (from Granger, 1963) exhibits the effect.

Sampling variance of the estimated spectral ordinate

8.14   Perhaps the most remarkable feature of the intensity, however, is the fact that the standard error of an ordinate does not diminish with n, but is equal to the ordinate itself. We saw that in the correlogram neighbouring ordinates were correlated but that the variances of individual serials were of order n^{-1}. By contrast, the neighbouring ordinates in a spectrum are virtually independent, but their variance is of order unity. "Neighbouring" here is a rather arbitrary term. In the correlogram neighbouring values are fixed by the interval of observation. In the intensity, theoretically, we have a continuous range of the variable α. In practice, of course, we compute the intensity for an equally spaced set of values from 0 to π. (Some machine programs have an option for the interval of calculation, and most of them can divide the range into at least 60 intervals.)

8.15   Consider the sums of $a(\alpha)$ and $b(\alpha)$ of (8.9) and (8.10), and in the first instance let u be a random series with zero mean and variance σ^2. It is easy to show that for large n and α not equal to 0 or π,

$$\frac{1}{n}\sum_{k=1}^{n}\cos^2\alpha\, k \to \frac{1}{2}, \quad \frac{1}{n}\sum_{k=1}^{n}\sin^2\alpha\, k \to \frac{1}{2}, \quad \frac{1}{n}\sum_{k=1}^{n}\cos\alpha\, k \sin\alpha\, k \to 0.$$

Table 8.2 *Residuals of logarithms of immigration data of Table 1.4, after removal of trend by a simple 17-point moving average. Values increased by 2*

Year	Value	Year	Value	Year	Value
1828	2·1395	1870	2·1981	1912	2·1482
	1·9973		2·1004		2·3036
1830	1·9754		2·1762		2·3413
	1·9039		2·1999		1·7889
	2·1679		2·0037		1·7653
	2·1922		1·8484		1·7853
	2·1800		1·7093		1·3817
	1·9737		1·6217		1·5165
	2·1627		1·6126		
	2·1435		1·7150	1920	2·0273
	1·7849	1880	2·1118		2·3278
	1·9688		2·2747		1·9536
1840	2·0015		2·3465		2·2463
	1·9354		2·2151		2·4339
	2·0087		2·1254		2·1182
	1·7894		1·9833		2·1919
	1·7860		1·8925		2·2630
	1·9087		2·0429		2·2600
	1·9958		2·0736		2·2739
	2·1363		2·0010	1930	2·2737
	2·0929	1890	2·0389		1·9115
	2·1826		2·1527		1·5264
1850	2·1702		2·1751		1·4049
	2·2743		2·0566		1·5707
	2·2451		1·8565		1·7101
	2·2375		1·7892		1·7898
	2·3080		1·8995		1·9839
	1·9764		1·7109		2·1388
	1·9792		1·6850		2·2386
	2·0760		1·7918	1940	2·1550
	1·7643	1900	1·9414		1·9775
	1·7562		1·9712		1·6609
1860	1·8668		2·0728		1·5275
	1·6451		2·1651		1·5560
	1·6444		2·1122		1·6421
	1·9340		2·1812		2·0608
	1·9564		2·1694		2·1608
	2·0432		2·2275		2·1900
	2·1462		2·0134		2·1938
	2·1266		2·0064	1950	2·2749
	2·0623	1910	2·1862		2·1353
1869	2·1682	1911	2·1510		2·1845
					1·9344
				1954	1·9700

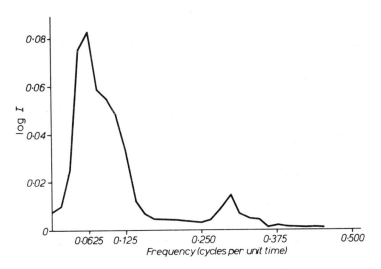

Fig. 8.3   Power spectrum of detrended U.S. immigration data of Table 1.4

If $\alpha = 0$ or π there is a discontinuity, the first expression tending to unity, the second to zero. It follows that if u_t in (8.9) and (8.10) is normal, $a(\alpha)$ and $b(\alpha)$ are distributed normally with zero mean and variance $\sigma^2/2\pi$; and that they are, moreover, independent. Hence

$$\frac{2\pi I}{\sigma^2} = \frac{2\pi(a^2 + b^2)}{\sigma^2} = 2\hat{w}(\alpha) \qquad (8.22)$$

is distributed as the sum of squares of two independent normal variables with unit variances, i.e. as χ^2 with two degrees of freedom. Now χ^2 with ν degrees of freedom has mean ν and variance 2ν. It follows that

$$E\{\hat{w}(\alpha)\} = 1 \qquad (8.23)$$

$$\text{var } \hat{w}(\alpha) = 1 = E^2\{w(\alpha)\}. \qquad (8.24)$$

8.16   This, as noted, applies to a random series. A theorem of Bartlett's (1955) shows that for weighted averages of independent random variables (and in particular for autoregressive or moving-average processes) it remains true, asymptotically, that

$$\text{var } w(\alpha) = E^2\{w(\alpha)\}. \qquad (8.25)$$

Let us summarize $a(\alpha)$ and $b(\alpha)$ in the single formula

$$J(\alpha) = a(\alpha) + ib(\alpha) = \frac{1}{\sqrt{(n\pi)}} \sum_{t=1}^{n} u_t\, e^{i\alpha t}. \qquad (8.26)$$

J has zero expectation. If the u's are a random series (not necessarily normal),

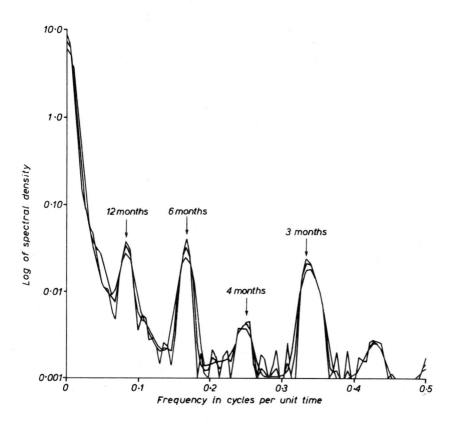

Fig. 8.4 Power spectrum of detrended U.S. bank clearing data (monthly, 1875-1958), showing peaking effects at 2-monthly, 3-monthly, etc. intervals (exponential linear trend removed)

$$E\{J(\alpha)J(\beta)\} = \frac{\sigma^2}{n\pi} \sum_{t=1}^{n} \exp\{i\alpha t + i\beta t\}$$

$$= \frac{\sigma^2}{n\pi} \frac{\exp(i\alpha + i\beta)[1 - \exp\{in(\alpha+\beta)\}]}{1 - \exp\{i\alpha + i\beta\}}. \tag{8.27}$$

If α and β are of the form $2\pi p/n$, p integral, the last term in the numerator vanishes. Then $J(\alpha)$ and $J(\beta)$ are uncorrelated. The same is true of the complementary $J^*(\alpha)$ defined by

$$J^*(\alpha) = a(\alpha) - ib(\alpha). \tag{8.28}$$

We know that $a(\alpha)$ and $b(\alpha)$ are uncorrelated. Likewise the ordinates in the spectrum at α and β are uncorrelated, provided that α, β are of the form $2\pi p/n$.

We have $I(\alpha) = J(\alpha)J^*(\alpha)$ and putting $\beta = -\alpha$ (8.27) we confirm the result that

$$E\{I(\alpha)\} = \sigma^2/\pi. \tag{8.29}$$

8.17   We have further

$$E\{I(\alpha)I(\beta)\} = \frac{1}{n^2\pi^2} E\left\{ \sum_{t=1}^{n} u_t e^{i\alpha t} \sum_{s=1}^{n} u_s e^{-i\alpha s} \sum_{k=1}^{n} u_k e^{i\beta k} \sum_{l=1}^{n} u_l e^{-i\beta l} \right\}$$

$$= \frac{1}{n^2\pi^2} \sum \{E(u_t u_s u_k u_l) \exp(\alpha t - \alpha s + \beta k - \beta l)\}. \tag{8.30}$$

For a random series the expectations vanish unless $t = s = k = l$, giving the fourth-order moment of u, or unless the suffixes are equal in pairs. If $t = s$ and $k = l$ the corresponding term is $E\{I(\alpha)\}E\{I(\beta)\}$. Hence after a little algebra we find

$$\text{cov}\{I(\alpha), I(\beta)\} = \frac{\kappa_4}{n\pi^2} + \frac{\sigma^4}{n^2\pi^2} \left\{ \frac{1 - \cos\{n(\alpha+\beta)\}}{1 - \cos(\alpha+\beta)} + \frac{1 - \cos\{n(\alpha-\beta)\}}{1 - \cos(\alpha-\beta)} \right\}. \tag{8.31}$$

For $\alpha = \beta$ we find, since $1 - \cos\theta = \frac{1}{2}\theta^2$ for θ small,

$$\text{var } I(\alpha) = \frac{\sigma^4}{\pi^2} + O(n^{-1}). \tag{8.32}$$

This confirms (8.25). Further, if u is non-normal the covariance is of order $1/n$. If it is normal the covariance is of order n^{-2} and vanishes if α, β are of the form $2\pi p/n$, p integral. In the latter case neighbouring ordinates are independent; even in the former the correlations are small.

8.18   Consider further the case when u_t is a weighted average of random variables, say

$$u_t = \sum_0^\infty g_s \, \epsilon_{t-s}. \tag{8.33}$$

We have, the suffix to J referring to the variable concerned,

$$J_u(\alpha) = \frac{1}{\sqrt{(n\pi)}} \sum_{s=0}^\infty \sum_{t=1}^n g_s \, \epsilon_{t-s} \, e^{i\alpha t}$$

$$= \frac{1}{\sqrt{(n\pi)}} \sum_{s=0}^\infty \sum_{t=1}^n \epsilon_{t-s} \, e^{i\alpha(t-s)} \, e^{i\alpha s} g_s$$

$$= \frac{1}{\sqrt{(n\pi)}} \sum_{k=0}^\infty \epsilon_{t-k} \, e^{i\alpha k} \sum_{s=0}^\infty e^{i\alpha s} g_s \quad \text{approximately}$$

$$= J_\epsilon(\alpha) T(\alpha), \tag{8.34}$$

where $T(\alpha)$ is the transfer function of g_s

$$T(\alpha) = \sum_{s=0}^\infty e^{i\alpha s} g_s. \tag{8.35}$$

We have at once
$$I_u(\alpha) = I_\epsilon(\alpha) T(\alpha) T^*(\alpha), \tag{8.36}$$

which is another form of the result obtained for the transfer function in the continuous case at equation (6.72). Further

$$\mathrm{E}\{I_u(\alpha)\} = T(\alpha) T^*(\alpha) \, \mathrm{E}\{I_\epsilon(\alpha)\} \tag{8.37}$$

and since the result is true for I_ϵ, we have asymptotically,

$$\mathrm{var}\, I_u(\alpha) = \mathrm{E}^2\{I_u(\alpha)\}. \tag{8.38}$$

The value is doubled at $\alpha = 0$ or π.

8.19 These various properties of the intensity, namely the side-band effects, echo effects, contributions from trend, and the uncontrolled fluctuation in the ordinates, obviously present serious problems in interpretation. Beveridge, for example, inspecting a diagram like Fig. 8.2 (although he worked with the periodogram before the dangers in this subject were realised), was led to suggest 19 periodicities in his series, whereas it is doubtful whether there are more than two, if that. The problem is dealt with by smoothing the spectrum. We find that in doing so we can reduce the sampling error to order $1/n$, but, as always, there is a price to pay: we bias the estimator.

8.20 We shall take what is equivalent to a moving average along the spectrum by means of a continuous weighting function h. In general the extent of this average will be the full length of the spectrum 2π, so in order to be able to apply it at all points we define it to be itself periodic, that is to say
$$h(u) = h(u + 2\pi). \tag{8.39}$$

Since it is a weighting function with unit total weight,
$$\int_{-\pi}^\pi h(u)\, du = 1. \tag{8.40}$$

The function is variously known as a "kernel" or "spectral window" (and not, for some obscure reason, as a moving average or a filter, although they are all the same thing). If $I(\alpha)$ is the intensity as computed from a sample, we construct the smoothed function

$$I_A(\alpha) = \int_{-\pi}^{\pi} h(u)I(\alpha - u)\,du$$

$$= \int_{-\pi}^{\pi} h(\alpha - u)I(u)\,du. \tag{8.41}$$

If $I(\alpha)$ is unbiased we see, on taking expectations, that $I_A(\alpha)$ in general will be biased, being an average of a number of different values only one of which is the true one.

8.21   In the choice of a suitable value of $h(u)$ we try to reconcile conflicting requirements. In order to avoid bias we should like the important part of h (i.e. that range for which its values are not small) to be narrow. But it must not be too narrow if it is to smooth effectively. A highly concentrated h can be regarded as possessing an effective range narrower than $-\pi$ to π, and this is sometimes known as the "bandwidth" of the function.

We may approximate to the integral (8.41) by the sum

$$I_A(\alpha) = \frac{2\pi}{n} \Sigma h(u_j)I(\alpha - u_j), \quad u_j = 2\pi j/n. \tag{8.42}$$

In practice, of course, this is the sum we should have to compute. In virtue of the choice of values of u_j and the remarks of **8.17** the values of I are uncorrelated for a finite set of values of α, and therefore

$$\text{var } I_A(\alpha) = \frac{4\pi^2}{n^2} \Sigma h^2(u_j) \text{ var } I(\alpha - u_j). \tag{8.43}$$

In virtue of (8.38) this is approximately

$$\doteq \frac{4\pi^2}{n^2} \Sigma h^2(u_j) \, \mathrm{E}\,\{I(\alpha - u)\}^2 \tag{8.44}$$

$$\doteq \frac{2\pi}{n} \int_{-\pi}^{\pi} h^2(u) \, \mathrm{E}\{I(\alpha - u)\}^2 \, du. \tag{8.45}$$

If $h(u)$ is concentrated in a narrow bandwidth we shall have

$$\text{var } I_A(\alpha) = \frac{2\pi}{n} \mathrm{E}\{I(\alpha)\}^2 \int_{-\pi}^{\pi} h^2(u)\,du. \tag{8.46}$$

As usual the value is twice as great at $\alpha = 0$ or π.

From (8.46) it follows that if the integral is bounded the variance of the averaged I is of order $1/n$. Here n is the number of points on the range 0 to 2π at which we measure the intensity (equation (8.42)).

It may similarly be shown that the correlation

$$\text{corr } \{I_A(\alpha), I_A(\beta)\} = \frac{\int_{-\pi}^{\pi} h(u)h(u - \alpha - \beta)\, du}{\int_{-\pi}^{\pi} h^2(u)\, du}. \tag{8.47}$$

8.22   From (8.46) it follows that for large n

$$\text{var } \log_A I(\alpha) \doteqdot \frac{\text{var } I_A(\alpha)}{I_A^2(\alpha)} = \text{Constant.} \tag{8.48}$$

This is one reason why the spectrum is sometimes drawn with ordinate equal to the logarithm of the intensity. The confidence intervals are then of constant width. Another reason is that the ordinates of the spectrum (as distinct from those of the periodogram) can be very large, and taking logarithms keeps the range of the ordinates within narrower limits.

Computation of ordinates

8.23   The actual computation of spectral ordinates can either proceed directly or through the intermediary of the serial correlations. The former is quicker, but the latter provides the correlogram as well. It also gives us a lead as to the appropriate choice of h functions. We have

$$I(\alpha) = \frac{1}{n\pi} \left\{ \left(\sum_{t=1}^{n} u_t \cos \alpha t \right)^2 + \left(\sum_{t=1}^{n} u_t \sin \alpha t \right)^2 \right\}$$

$$= \frac{1}{n\pi} \sum_{s,t=1}^{n} u_s u_t (\cos \alpha t \cos \alpha s + \sin \alpha t \sin \alpha s)$$

$$= \frac{1}{\pi} \sum_{k=-n+1}^{n-1} c_k \cos k\alpha, \tag{8.49}$$

where c_k is a covariance-type expression defined by

$$c_k = \frac{1}{n} \sum_{t=1}^{n-k} u_t u_{t+k}, \quad k > 0. \tag{8.50}$$

For infinite n the mean of (8.49) reduces to the known expression

$$E\{I(\alpha)\} = \frac{\sigma^2}{\pi} w(\alpha) = \frac{\sigma^2}{\pi} \sum_{-\infty}^{\infty} \rho_k \cos k\alpha. \tag{8.51}$$

For large n the sum in (8.49) is extensive. Let us consider the effect of truncating it at some value $k = \pm q$, so defining

$$E\{I_q(\alpha)\} = \frac{1}{n} \sum_{-q}^{q} \lambda_k c_k \cos k\alpha, \quad \lambda_k = \lambda_{-k}. \tag{8.52}$$

The λ's are constants, left for the moment undetermined, by the choice of which we may be able to improve the estimates. In parental form (8.52) is equivalent to

$$I_q(\alpha) = \frac{\sigma^2}{\pi} \sum_{-q}^{q} \lambda_k \rho_k \cos k\alpha \qquad (8.53)$$

or to

$$w_q(\alpha) = \sum_{-q}^{q} \lambda_k \rho_k \cos k\alpha. \qquad (8.54)$$

Now from equation (6.54),

$$\rho_k = \frac{1}{\pi} \int_0^\pi w(\alpha) \cos k\alpha \, d\alpha.$$

On substitution in (8.54) we find that

$$w_q(\alpha) = \frac{1}{\pi} \sum_{-q}^{q} \lambda_k \int_0^\pi w(u) \cos ku \cos k\alpha \, du$$

$$= \int_{-\pi}^{\pi} w(u) \left\{ \frac{1}{2\pi} \sum_{-q}^{q} \lambda_k \cos ku \cos k\alpha \right\} du. \qquad (8.55)$$

The use of the truncating formula (8.52) then turns out, rather unexpectedly perhaps, to be equivalent to taking as a weighting function

$$h(\beta) = \frac{1}{2\pi} \sum_{-q}^{q} \lambda_k \cos k\beta \cos k\alpha. \qquad (8.56)$$

Provided that $\lambda_0 = 1$, this is readily verified to satisfy (8.39) and (8.40). We also find

$$\int_{-\pi}^{\pi} h^2(\beta) \, d\beta = \frac{1}{2\pi} \sum_{-q}^{q} \lambda_k^2 \cos^2 k\alpha. \qquad (8.57)$$

8.24   A number of different h functions have been proposed by different writers. We may notice four in particular, the last three being those most frequently met with in machine programs.

(1)   (Daniell, 1946—the first of its kind)
Take

$$\lambda_k = \frac{\sin kl}{kl}, \quad l > 0. \qquad (8.58)$$

In virtue of the known integrals

$$\int_0^\infty \frac{\sin px \cos qx}{x} \, dx = \tfrac{1}{2}\pi, \quad |p| > |q|,$$
$$= \tfrac{1}{4}\pi, \quad |p| = |q|,$$
$$= 0, \quad |p| < |q|, \qquad (8.59)$$

the weighting function will be found to be approximated by the integral

$$\frac{1}{2\pi} \int_0^\infty \frac{\sin lx}{lx} [\cos \{\tfrac{1}{2}(\alpha + \beta)x\} + \cos \{\tfrac{1}{2}(\alpha - \beta)x\}] \, dx$$

$$= \frac{1}{2l}, \quad -l \leqslant \alpha - \beta \leqslant l$$

$$= 0, \quad \text{otherwise.} \tag{8.60}$$

(2)   Bartlett (1950)
   Take

$$\lambda_k = 1 - \frac{k}{q}.$$

We find

$$h(\beta) = \frac{1}{\pi} \left\{ \frac{\sin^2 \{\tfrac{1}{2}q(\alpha + \beta)\}}{q \sin^2 \{\tfrac{1}{2}(\alpha + \beta)\}} + \frac{\sin^2 \{\tfrac{1}{2}q(\alpha - \beta)\}}{q \sin^2 \{\tfrac{1}{2}(\alpha - \beta)\}} \right\}. \tag{8.61}$$

(3)   Blackman–Tukey (1958)
   Take

$$\lambda_k = 1 - 2a - 2a \cos\left(\frac{\pi k}{q}\right).$$

We find

$$h(\gamma) = \frac{1}{\pi} \left[(1 - 2a) \frac{\sin(q + \tfrac{1}{2})\gamma}{\sin\tfrac{1}{2}\gamma} + a \left\{ \frac{\sin\{(q + \tfrac{1}{2})\gamma + \pi/q\}}{\sin(\tfrac{1}{2}\gamma + \pi/q)} + \frac{\sin\{(q + \tfrac{1}{2})\gamma - \pi/q\}}{\sin\{\tfrac{1}{2}\gamma - \pi/q\}} \right\} \cdots \right] \tag{8.62}$$

where $\gamma = \alpha - \beta$, together with a similar term obtained by putting $\gamma = \alpha + \beta$.

(4)   Parzen (1961)
   Take

$$\lambda_k = 1 - 6\left(\frac{k}{q}\right)^2 + 6\left(\frac{k}{q}\right)^3, \quad 0 \leqslant k \leqslant \tfrac{1}{2}q,$$

$$= 2\left(1 - \frac{k}{q}\right)^3, \quad \tfrac{1}{2}q \leqslant k \leqslant q.$$

We find

$$h = \frac{3}{4\pi q^3} \left\{ \frac{\sin\tfrac{1}{4}\gamma q}{\sin\tfrac{1}{4}\gamma} \right\}^4. \tag{8.63}$$

In Fig. 8.1 and 8.2 (pages 100 and 102) we have drawn the smoothing achieved by the use of a Parzen function (8.63).

8.25   There has been a great deal of literature on spectrum analysis published since 1950, distinguished, perhaps, more by its ingenuity and mathematical expertise than its practical applicability. We may notice as possible further reading the papers by Parzen (e.g. 1961), Daniels (1962), and those by Hannan (1960b) and Durbin (1961) concerning the effect of seasonal variation on the spectrum. Among books on the subject should be mentioned Blackman and Tukey (1958), Grenander and Rosenblatt (1957), Granger and Hatanaka (1964) on economic series, and the symposium edited by Rosenblatt (1963).

8.26   As with the correlogram, one of the primary problems of the spectrum is to know what interpretation to put upon it. It seems that both correlogram and spectrum are very useful in suggesting what kind of models could account for the data, or perhaps, equally usefully, in dismissing some models as inadequate. In either case, as it seems to me, we should consider the diagrams as a whole, which is one reason why we do not, in practice, need to spend much time on testing the significance of individual serial correlations or ordinates in the spectrum.

Seasonality and harmonic components

8.27   We can now pick up a point which had to be left on one side in Chapter 5, the treatment of seasonal effects by harmonic analysis. This will, at the same time, help to illustrate the use of transfer functions in gauging the effect of moving averages on different components of the series.

For a moving average of extent $2m + 1$, which we can conveniently represent with symmetric weights a_{-m} to a_m, the transfer function reduces to the real quantity

$$T(\alpha) = a_0 + 2 \sum_{j=1}^{m} a_j \cos j\alpha. \qquad (8.64)$$

The effect of taking such an average is to multiply the spectral ordinates by the square of the modulus of $T(\alpha)$. Given the weights, we can compute T for a range of values from 0 to π and hence determine the effect on the "filtered" series. Consider in particular the transfer functions of a centred moving average of 12 and a Spencer 15-point. The following Table 8.3 (Burman, 1965) shows the transfer functions. Here, for example, with an angular frequency of $45°$ the function for the centred moving average is

$$\tfrac{1}{24}[2 + 2\{\cos 45° + 2 \cos 90° + 2 \cos 135° + 2 \cos 180° + 2 \cos 225° + \cos 270°\}]$$
$$= \tfrac{1}{24}\{2 + 2(-2 - \sqrt{2})\} = -0.201.$$

The values of the transfer function of the centred average fall to 10 per cent. or less after an angular frequency of $45°$, corresponding to cycles per unit (month) of $\tfrac{1}{8}$ or a wavelength of 8 months. For cycles of shorter wavelength the spectrum of the series after trend removal is not greatly affected; for larger cycles the effect is substantial and they would be mostly removed with the trend, if any. The Spencer 15-point is of the same kind and on the whole distorts the residual spectrum rather less. The general effect of either is to remove with the trend some or all of the long cycles, but to leave the shorter ones untouched for the most part.

Table 8.3 *Transfer functions of a centred 12-point average and a Spencer 15-point average (Burman, 1965)*

Angular frequency (degrees)	T.F. Centred 12-point	T.F. Spencer 15-point	Angular frequency (degrees)	T.F. Centred 12-point	T.F. Spencer 15-point
0	1·000	1·000	95	−0·038	−0·002
5	0·955	1·000	100	− ·061	− ·007
10	·824	1·003	105	− ·064	− ·012
15	·633	0·984	110	− ·051	− ·015
20	·409	·952	115	− ·027	− ·016
25	·188	·895	120	0	− ·013
30	0	·809	125	·022	− ·008
35	− ·133	·696	130	·034	− ·003
40	− ·198	·564	135	·034	0
45	− ·201	·425	140	·026	·001
50	− ·155	·293	145	·013	0
55	− ·080	·180	150	0	− ·003
60	0	·094	155	− ·009	− ·005
65	·065	·037	160	− ·013	− ·005
70	·103	·006	165	− ·011	− ·004
75	·109	− ·005	170	− ·006	− ·003
80	·086	− ·005	175	− ·002	− ·001
85	·045	− ·002	180	0	0
90	0	0			

8.28 With a suitable trend-eliminator, therefore, we can proceed to the determination of the seasonal component secure in the knowledge that the higher frequencies are only slightly affected. Moreover, since by definition seasonality is strictly periodic, we expect to be able to represent it by a sum of harmonics in the Fourier manner, namely with regular frequencies $\alpha_k = 2\pi k/12$ and wavelengths 12, 6, 4, 3, 2·4 and 2 months. Thus, if x_t is the deviation from trend of the tth month in a set of $12p$ months, we have

$$a_j = \frac{2}{12p} \sum_{t=1}^{12p} x_t \cos \alpha_j t, \quad j = 1, 2, \ldots, 5 \qquad (8.65)$$

$$b_j = \frac{2}{12p} \sum_{t=1}^{12p} x_t \sin \alpha_j t, \quad j = 1, 2, \ldots, 5 \qquad (8.66)$$

$$a_6 = \frac{1}{12p} \sum_{t=1}^{12p} x_t \cos \alpha_6 t. \qquad (8.67)$$

The seasonal movement will then be represented by a set of 11 constants.

8.29 Having determined the seasonal movement in this way we can, if necessary, find its spectrum, subtract from the transferred spectrum, and "restore" the spectrum of the deseasonalized series by multiplying it by $1 - T(\alpha)$. A slight modification enables us to take account of moving seasonal effects. We can, in fact, analyse the series year by year, taking $p = 1$ in (8.65), and hence obtain 11 annual series of the 11 constants. Each of these series can be smoothed and extrapolated and the resulting values used to estimate the seasonal component for any given year. Since the Fourier constants are un-correlated, the smoothing can proceed independently for each series. Burman (1965, 1966) has used a method based on the foregoing ideas to adjust some financial series. He takes the model as additive and devised a 13-point aver-age instead of the Spencer 15-point. His projections over the last values of the series, where the symmetric moving average leaves a gap, are based on exponential forecasting methods which we discuss in a later chapter.

8.30   Spectral methods have also been used (U.S. Bureau of the Census, 1965) to discuss the adequacy of methods of seasonal elimination. It has been known for some time that the spectrum of an economic series very often exhibits a J-shaped form, there being high values at low frequencies (cf. Granger and Hatanaka, 1964). If seasonality is removed from such a series, e.g. by the Census X–11 method already mentioned, and the program overdoes it by removing too much, these high values will show up on the spectrum fluctuations at the 6-month, 4-month, etc. points.

8.31   Whether seasonal effects are estimated by the methods of Chapter 5 or by the use of spectrum analysis, it seems preferable to isolate them before proceeding with further analysis of the series. Some of the autoprojective methods which we describe below leave the seasonal element in the series which is to be forecast. Further study may elucidate the situation to some extent; at the moment the choice is very much a matter of personal preference.

NOTES

(1) The Beveridge data on wheat prices have recently been re-examined by Granger and Hughes (1971) who conclude that they do contain a cycle of about 13·3 years. Kendall (1945) and Sargan (1953) regarded the series as adequately represented by a Yule autoregressive scheme.

(2) Preferences for the type of "window" employed to smooth the spectrum vary, but the Parzen version has the advantage that it does not lead to negative ordinates, whereas some of the others may do so.

(3) Although the spectrum and the correlogram determine each other unique-ly, neither determines the original series.

9

Forecasting by autoprojective methods

9.1 The English language is rich in words which connote attempts to see into the future: forecasting, foretelling, foreseeing (even foretasting); prediction and prevision; prognostication; not to mention phrases such as "crystal-gazing". There are two which are used to denote numerical forecasting methods, namely forecasting and prediction. The two are often used synonymously. If there is any difference, forecasting is perhaps used in reference to specific events of a quantifiable kind, prediction to more verbal descriptions relating to ambient circumstances — most of us would speak of forecasting the results of a general election but of predicting its effect on foreign policy. For the purposes of this chapter we shall not try to draw fine distinctions between the two terms.

9.2 The methods we use in forecasting are not independent either of the time-span over which we are looking or of the purposes to which the forecasts are to be put. Broadly speaking, we may delimit two cases: (1) we may require a forecast which is to be the basis of our own action, as in sales forecasting; (2) an agency may produce a forecast which is going to be used for all kinds of purposes by different people, as when someone forecasts the population, or a Government agency forecasts growth rates in sectors of the economy. The prudent forecaster (and in forecasting one needs to be prudent, because mistakes are remembered much longer than successes) will regard his range of techniques as a mechanic regards his tool-bag, using whichever instrument the circumstances require.

9.3 It is customary to divide the subject into three parts: short-term, medium-term and long-term forecasting. These are terms relative to the subject under study. In meteorology, for example, "short-term" may mean only two or three days ahead, and "medium" refers to the next few months; whereas in economics "short-term" means a few months, perhaps as much as a year, "medium" usually refers to the next five years, and everything after that is "long-term". The periods concerned are very often determined by extraneous factors: by month for sales, by year for accountancy (at least sometimes), by quinquennia for Universities or British electoral parties, and so on.

9.4   We may dismiss at the outset the vaguer, almost literary, subjects of long-term forecasting and technological forecasting. Undoubtedly they are useful subjects to consider and discuss, but there is not much point in trying to quantify except in the broadest terms. A few exceptions to this rule exist; for example, the forecasting of human populations over the next thirty years, although even here it would be hazardous to lean too heavily on numerical estimates. Such forecasting is not amenable to sophisticated mathematical or statistical techniques; some would think that it was not amenable to any techniques, sophisticated or not.

9.5   We shall therefore concentrate on medium and short-term forecasting. Before proceeding to the more advanced methods, however, we may recall that in some cases forecasting can be successfully carried out merely by watching the phenomenon of interest approach. This is how we forecast the weather for a particular spot — we set up stations to watch it coming. We can forecast demands on our educational system over the next twenty years with some confidence because a good many of the young men and women who will be exposed to it are already born. Nor should one despise these simple-minded methods in the behavioural sciences — exports next month can be predicted from the order book for this and previous months; and it has been claimed that good forecasts of next year's consumption of consumer durables can be obtained by surveys conducted now to ask people what they intend to buy.

Autoprojection

9.6   The first method to consider is that for which we extrapolate forwards an existing series, without regard to other series which may be concomitant. There is little to be said about predictive decomposition other than what is implicit in the decomposition itself: we analyse into trend, short-term oscil-lation, seasonal effect, and random residual, and project the first three ele-ments forward, reassembling them to form the forecast, by addition or multi-plication as the case may be. Only one point remains for examination, namely the errors of forecast.

9.7   In ordinary statistics we are accustomed to compute standard errors of estimate or to put confidence intervals around an estimate, expressing in probabilistic terms the degree of assurance that it lies in a certain range. All this is predicated on the assumption that the sample on which we are working was randomly chosen from a parent entity.

The straightforward application of such procedures in forecasting is not legitimate in general, whatever has been said to the contrary in the literature. In performing a predictive decomposition we have set up a model. Any errors of forecast will be due not merely to sampling effects in the customary sense, but to mistakes in the specification of that model. We can, of course, legitimately ask questions such as: if the trend is really represented by a cubic curve, what are the errors we are likely to encounter in virtue of the fact that we have fitted a cubic to the data? But this does not answer the question as to what the errors will be if we should have fitted a quartic. It will be evident from the recursive nature of the procedure in the Census Mark II program, for example, that these might have been mistakes of specification at various points along the way. It is hard to put a standard error on the results produced by such a program — indeed the concept of standard error needs some re-examination.

9.8   The safest way to gauge the reliability of a forecasting method is to consider its performance over a period. From a suite of observed errors we can form a good empirical estimate of the error likely to be encountered in the future. Unfortunately, a strict interpretation of this rule would involve more work that we can usually concede to it, because, if we go back into the past and examine what the errors would have been, we ought to perform the forecast anew at each past time-point and make it only on the basis of what we actually knew at that point. This is usually much too tedious, and in one sense it does not really answer the question about the future, because we may by now have accumulated enough experience to improve on the errors committed some time ago. It appears to be a reasonable compromise to calculate on the basis of current knowledge what the errors would have been. This may underestimate the true errors, but at least it gives us some guide to errors of the future.

9.9   In discussing the weights involved in trend fitting at the start of a series we noted **(3.10)** that the values given at the extremes of a polynomial are much less reliable than those at the beginning. This is generally true, and indeed the errors of estimation as we proceed to extrapolate the polynomial *beyond* the range of fitting increase quite rapidly.

Example 9.1

Consider the very simple case of fitting a straight line to three points at time-points $-1, 0, 1$. The line is

$$y = \bar{u} + \tfrac{1}{2}(-u_{-1} + u_1)t$$
$$= \tfrac{1}{6}\{u_{-1}(2 - 3t) + 2u_0 + (2 + 3t)u_1\}. \tag{9.1}$$

Applied to a random series with variance ϵ, this gives a variance around the true linear value of

$$\text{var } \epsilon \cdot \tfrac{1}{36}\{(2-3t)^2 + 2^2 + (2+3t)^2\} = \tfrac{1}{6}(2+3t^2) \text{ var } \epsilon. \qquad (9.2)$$

Thus for $t = 1, 2, 3, 4$, etc., the error-reducing power varies as the square of t.

Exponential smoothing

9.10   The fitting of systematic functions to observation, by whatever method, has up to this point been based on least-squares criteria in which all the observations are of equal weight. It may be felt that, in some sense, more weight should be given to the recent past and that observations taken a long time ago should be discounted in comparison. To a certain extent we allowed for this in moving averages of finite extent, where the values ascribed to the recent block of $2m + 1$ values are not dependent on previous values. We now consider a rather different method of emphasizing the more recent observations.

Consider a set of weights proportional to powers of a factor β, namely 1, β, β^2, β^3, etc. Since their sum must be unity, and

$$\sum_{j=0}^{\infty} \beta^j = \frac{1}{1-\beta}, \qquad (9.3)$$

the actual weights are $(1-\beta), (1-\beta)\beta, (1-\beta)\beta^2$, etc. We suppose that $|\beta| < 1$. Consider also a process which is a constant α_0 plus a random residual ϵ with mean zero. We construct the predictor at time t:

$$a_0(t) = (1-\beta)\{x_t + \beta x_{t-1} + \beta^2 x_{t-2} + \ldots\} = (1-\beta)\sum_{j=0}^{\infty} \beta^j x_{t-j}. \qquad (9.4)$$

Here, owing to the fact that $|\beta| < 1$, the weights diminish exponentially. Writing $t-1$ for t, we have

$$a_0(t-1) = (1-\beta)\{x_{t-1} + \beta x_{t-2} + \ldots\}. \qquad (9.5)$$

Hence

$$a_0(t) = (1-\beta)x_t + \beta a_0(t-1). \qquad (9.6)$$

The formula bears some resemblance to a Markoff process, but x_t, in the role of the residual, is not independent from one time-period to the next.

If, then, we know the constant β, we can construct an estimator of $a_0(t)$ at time t from the estimate made at time $t-1$ plus $(1-\beta)$ times the actual observation at time t. This could provide an estimate of the future values of the series.

9.11   Let the error made at time $t-1$ in forecasting x_t one unit ahead be e_t, that is to say

$$x_t - a_0(t-1) = e_t. \qquad (9.7)$$

Substituting for $a_0(t-1)$ in (9.6), we get

$$a_0(t) = x_t - \beta e_t = a_0(t-1) + (1-\beta)e_t. \tag{9.8}$$

Thus our forecast at time t is the forecast made at time $t - 1$ increased by $(1 - \beta)$ times the undershoot last time.

The formula (9.3) from which we began is useless of itself, for it explicitly contains all the past observations. However, owing to the structure of the estimate we can make a forecast at any one time based solely on information from the last time-period together with the current observations.

9.12 The method thus has two valuable features. Once β is determined, it does not require a knowledge of the past beyond one time-unit ago; and it is a quick method, easily updated from one time-point to the next. It will be seen from (9.8) that if the errors are fairly small the forecasts are relatively insensitive to variations in the magnitude of β. There is, of course, one sense in which we require all the past data, namely in the estimation of β; but it is customary to perform that estimation at very infrequent intervals, not every time a forecast is required.

The actual estimation is not very easy, and requires a routine for an electronic computer. We have, in fact, to minimize for β the sum of squares of errors

$$\sum_{-\infty}^{t_0} \{x_t - (1-\beta) \sum_{j=1}^{\infty} \beta^j x_{t-j}\}^2. \tag{9.9}$$

To differentiate this with respect to β yields a most unwieldly equation for β. In practice we truncate the sum $\Sigma \beta^j x_{t-j}$ at some point $t = k$, by which β^j may be considered small, and minimize by successive approximation. A repetition of the computation with $t = k + 1$, and if necessary $k + 2$, etc., will show whether a reasonable approximation to β has been reached. Great accuracy, as we have remarked, is not necessary here.

9.13 We can also regard a_0 as determined by minimizing the *weighted* sum of squares

$$\sum_{j=0}^{\infty} (x_{t-j} - \alpha_0)^2 \beta^j. \tag{9.10}$$

Differentiation with respect to β leads to the estimator of (9.4), which may then be regarded as optimal in a rather limited sense.

9.14 The above is the simplest case of what is sometimes known as "adaptive" forecasting. The precise meaning of "adaptive" is not clear. In a sense every method of forecasting is adaptive if it brings into account any observations made since the last forecast. Nor is β "adapted" except perhaps occasionally. The general understanding of the term seems to be that "adaptive" forecasting allows the updating of forecasts with a minimum of delay and arithmetical nuisance, e.g. by equation (9.8).

Example 9.2

Assume that $\beta = 0.6$ and that our estimate of the sales of a certain product, made one month ago, was 120. If in fact it turns out to have been 130, the estimate we now make is

$$120 + (1 - 0.6)\{130 - 120\} = 124.$$

Since we are estimating a constant, this figure of 124 could be used to forecast, not merely one step ahead, but several. This is due to the very simple model we have assumed. In general, we shall have different forecasts for different periods ahead.

9.15   We now extend the method to the case where the true value is given by a linear function $\alpha_0 + \alpha_1 t$. As before, for given β, we minimize

$$\sum_{j=0}^{\infty} \{x_{t-j} - \alpha_0 + \alpha_1 j\}^2 \beta^j, \tag{9.11}$$

leading to the estimators

$$(1 - \beta) \Sigma x_{t-j} \beta^j - a_0 + (1 - \beta) \Sigma j \beta^j a_1 = 0 \tag{9.12}$$

$$\Sigma x_{t-j} j \beta^j - a_0 \Sigma j \beta^j + a_1 \Sigma j^2 \beta^j = 0. \tag{9.13}$$

Now $\Sigma j \beta^j = \beta/(1 - \beta)^2, \quad \Sigma j^2 \beta^j = \beta(1 + \beta)/(1 - \beta)^3.$ \tag{9.14}

The estimating equations reduce to

$$(1 - \beta) \Sigma x_{t-j} \beta^j - a_0 + a_1 \frac{\beta}{1 - \beta} = 0 \tag{9.15}$$

$$(1 - \beta)^2 \Sigma x_{t-j} j \beta^j - \beta a_0 + a_1 \frac{\beta(1 + \beta)}{1 - \beta} = 0. \tag{9.16}$$

Writing

$$S_1(t) = (1 - \beta) \sum_{j=0}^{\infty} x_{t-j} \beta^j,$$

which may be regarded as a first smoothing of x_t from x_t backwards, let us apply a second smoothing

$$\begin{aligned}
S_2(t) &= (1 - \beta)\{S_1(t) + \beta S_1(t - 1) + \beta^2 S_1(t - 2) + \ldots\} \\
&= (1 - \beta)^2 \{x_t + \beta x_{t-1} + \beta^2 x_{t-2} + \ldots \\
&\qquad\qquad + \beta x_{t-1} + \beta^2 x_{t-2} + \ldots \\
&\qquad\qquad\qquad\qquad + \beta^2 x_{t-2} + \ldots\} \\
&= (1 - \beta)^2 \Sigma x_{t-j} \beta^j (j + 1) = (1 - \beta)^2 \Sigma x_{t-j} j \beta^j + (1 - \beta) S_1. \tag{9.17}
\end{aligned}$$

We may then write (9.15) and (9.16) as

$$S_1 - a_0 + \frac{\beta}{1 - \beta} a_1 = 0 \tag{9.18}$$

$$S_2 - (1-\beta)S_1 - \beta a_0 + \frac{\beta(1+\beta)}{1-\beta}a_1 = 0, \qquad (9.19)$$

giving

$$S_1 = a_0 - \frac{\beta}{1-\beta}a_1 \qquad (9.20)$$

$$S_2 = a_0 - \frac{2\beta}{1-\beta}a_1, \qquad (9.21)$$

or alternatively

$$a_0 = 2S_1 - S_2 \qquad (9.22)$$

$$a_1 = -\frac{1-\beta}{\beta}(S_2 - S_1). \qquad (9.23)$$

9.16 We now require to be able to express the values of a_0 and a_1 at time t in terms of those at time $t-1$. We know that, as at (9.6),

$$S_1(t) = (1-\beta)x_t + \beta S_1(t-1) \qquad (9.24)$$

and by the same argument

$$S_2(t) = (1-\beta)S_1(t) + \beta S_2(t-1) \qquad (9.25)$$
$$= (1-\beta)^2 x_t + \beta(1-\beta)S_1(t-1) + \beta S_2(t-1). \qquad (9.26)$$

Hence from (9.20) to (9.23) we find

$$a_0(t) = (1-\beta^2)x_t + \beta^2 a_0(t-1) + \beta^2 a_1(t-1) \qquad (9.27)$$
$$a_1(t) = (1-\beta)^2 x_t - (1-\beta)^2 a_1(t-1) - (1-\beta)^2 a_0(t-1) + a_1(t-1). \qquad (9.28)$$

Writing the error of estimation one step ahead as

$$e_t = x_t - a_0(t-1) - a_1(t-1), \qquad (9.29)$$

we have

$$a_0(t) = a_0(t-1) + a_1(t-1) + (1-\beta^2)e_t \qquad (9.30)$$
$$a_1(t) = a_1(t-1) + (1-\beta)^2 e_t. \qquad (9.31)$$

These are the formulae which enable us to update the estimators on the basis of last year's estimates and the error. Our forecast one unit ahead will be

$$a_0(t) + a_1(t) = a_0(t-1) + 2a_1(t-1) - 2\beta(1-\beta)e_t. \qquad (9.32)$$

9.17 The foregoing procedure, which is essentially due to Brown (1963), can be generalized to trends of higher order in n with nothing more than algebraical complexity. For example, with a quadratic trend $\alpha_0 + \alpha_1 t + \alpha_2 t^2$, the updating formulae are

$$a_0(t) = a_0(t-1) + a_1(t-1) + a_2(t-1) + (1-\beta^3)e_t \qquad (9.34)$$
$$a_1(t) = a_1(t-1) + 2a_2(t-1) + \tfrac{3}{2}(1-\beta)(1-\beta^2)e_t \qquad (9.35)$$
$$a_2(t) = a_2(t-1) + \tfrac{1}{2}(1-\beta)^3 e_t. \qquad (9.36)$$

Holt–Winters model

9.18 It is, of course, fairly to be questioned whether the restriction of such a model to a single parameter β is not too drastic. Holt (1957), the first to consider these simple updating type models, in fact admitted two parameters. The method was extended by Winters (1960) to cover seasonal effects, and the Holt–Winters model therefore includes three parameters. The forecast (made at time t) k units ahead is, for a linear trend,

$$\{a_0(t) + a_1(t)k\}s(t + k - L), \tag{9.37}$$

where s is the seasonal factor and L is the number of points in the year at which the series is observed (e.g. 12 for monthly data). The formula is then multiplicative in seasonality. The updating formulae, which we quote without proof, are

$$a_0(t) = \beta_1 \frac{x_t}{s(t - L)} + [(1 - \beta_1)\{a_0(t - 1) + a_1(t - 1)\}] \tag{9.38}$$

$$s(t) = \beta_2 \frac{x_t}{a_0(t)} + (1 - \beta_2)s(t - L) \tag{9.39}$$

$$a_1(t) = \beta_3\{a_0(t) - a_0(t - 1)\} + (1 - \beta_3)a_1(t - 1), \tag{9.40}$$

where $\beta_1, \beta_2, \beta_3$ are the parameters.

9.19 One of the practical difficulties in fitting such a model is to estimate the constants. The usual practice is to take a range of possible values, and choose among them by calculating the sum of squares of residuals for each set.

Harrison's seasonal model

9.20 Harrison (1965) modified the Holt–Winters model by expressing the seasonal element in terms of harmonics, similar to Burman's treatment of seasonals referred to in **8.27**. We define

$$a_k = \frac{2}{L} \sum_{j=1}^{L} s(t - L + j) \cos k\lambda_j \tag{9.41}$$

$$b_k = \frac{2}{L} \sum_{j=1}^{L} s(t - L + j) \sin k\lambda_j, \tag{9.42}$$

where
$$\lambda_j = \frac{2(j - 1)\pi}{L} - \pi \tag{9.43}$$

and the smoothed seasonal factors are

$$s(t - L + j) = 1 + \sum_k \{a_k \cos (k\lambda_j) + b_k \sin (k\lambda_j)\}, \tag{9.44}$$

where the summation on the right covers those harmonics which are judged to be significant. Harrison then develops updating formulae for which reference may be made to his 1965 paper.

Box–Jenkins model

9.21 Box and Jenkins (1970) reverted to a method based on an autoregressive series, in which no prior assumptions are made about discounting factors. They deal with the problem of trend by working on the differences of the series, and allow the residuals to be correlated by regarding them as a finite moving average of random ϵ's. Thus, writing ∇ for the backward difference,

$$\nabla u_t = u_t - u_{t-1}. \tag{9.45}$$

Box and Jenkins consider the (non-seasonal) model

$$\nabla^d u_t - \alpha_1 \nabla^d u_{t-1} - \alpha_2 \nabla^d u_{t-2} - \ldots - \alpha_p \nabla^d u_{t-p}$$
$$= \epsilon_t + \beta_1 \epsilon_{t-1} + \ldots + \beta_q \epsilon_{t-q}. \tag{9.46}$$

If B is the backward shift operator defined by $Bu_t = u_{t-1}$, so that $\nabla = 1 - B$, this can be written

$$\alpha(B)(1 - B)^d u_t = \beta(B)\epsilon_t, \tag{9.47}$$

where $\alpha(B)$ and $\beta(B)$ are polynomials in B of order p and q.

9.22 As usual, one of the main problems in applying such a model is to find efficient estimates of the parameters involved. We have here three types, namely the order of difference d, the p autoregressive parameters α, and the q moving-average parameters β. In practice it appears that none of p, d, q is required to exceed 2, but even so the problem of estimation is rather involved. For full details reference may be made to Box–Jenkins' book, but broadly speaking the procedure is as follows: we first difference the series until it appears to be stationary in mean and variance, and hence have an estimate of d. (If there is any doubt, we may have to consider several neighbouring values of d.) The problem then reduces to estimating the constants in an autoregressive moving-average model.

$$\alpha(B)v_t = \beta(B)\epsilon_t, \tag{9.48}$$

where $v(t)$ is the dth difference of the original series. The ϵ's on the right in (9.48) do not appear in v_{t-q-1}. If we multiply both sides by v_{t-q-j} ($j \geqslant 1$) and take expectations, the right-hand side vanishes. Writing c_k for the serial covariance of order k, we then have

$$c_{q+1} - \alpha_1 c_q - \alpha_2 c_{q-1} - \ldots - \alpha_p c_{q-p+1} = 0 \tag{9.49}$$

and $(p - 1)$ further equations by multiplying v_t by v_{t-q-2}, etc. The solution of these equations gives us first estimates of the α's (unfortunately, owing to the sampling variation of the higher serial covariances, not very reliable estimates). With these estimates we can then determine the left-hand side of equation (9.48), and the serial correlations of the results give us equations in the β's – themselves, unfortunately, quadratic. The solution of these equations gives us first estimates of the β's. These preliminary estimates of α and β are then used as the starting point of a machine routine which re-estimates by minimizing the sum of squares of residuals, i.e. the left- less the right-hand side in (9.48).

Further research may result in improved methods of procedure. Compare **12.8** and **12.9**.

9.23　Defining now a range of coefficients ψ_j by

$$\alpha(B)(1-B)^d\{1-\psi_1 B-\psi_2 B^2+\ldots\} = \beta(B), \qquad (9.50)$$

we see that

$$u_{t+k} = \sum_{j=0}^{\infty} \psi_j \epsilon(t+k-j). \qquad (9.51)$$

Likewise, defining coefficients π_j by

$$\alpha(B)(1-B)^d = \{1-\pi_1 B-\pi_2 B^2-\pi_3 B^3-\ldots\}\beta(B), \qquad (9.52)$$

we have

$$u_{t+k} = \sum_{j=1}^{\infty} \pi_j u_{t+k-j} + \epsilon_{t+k}. \qquad (9.53)$$

Given p, d, q, α, β we can determine the ψ's and π's without difficulty. In general the π's diminish to zero fairly rapidly, and indeed it is a condition for the validity of the Box—Jenkins method that the π's form a convergent series. Equation (9.53) is therefore the one favoured for forecasting. Let the forecast (made at time t) k units ahead be, as at (9.51), a linear function of the residuals:

$$F_k(t) = p_k \epsilon_t + p_{k+1}\epsilon_{t-1} + p_{k+2}\epsilon_{t-2} + \ldots, \qquad (9.54)$$

where, for the moment, we will not assume anything about the coefficient p. Then the mean-square error of forecast is

$$E\{v_{t+k}-F_k(t)\}^2 = E\{\sum_{j=0}^{\infty} \psi_j \epsilon_{t+k-j} - \sum_{j=0}^{\infty} \epsilon_{t-j} p_{k+j}\}^2$$

$$= \text{var } \epsilon\{1+\psi_1^2+\ldots+\psi_{k-1}^2 + \sum_{j=0}^{\infty} (\psi_{j+k}-p_{j+k})^2\}. \qquad (9.55)$$

For this to be minimized $\psi_{j+k} = p_{j+k}$. Then

$$F_k(t) = \sum_{j=0}^{\infty} \psi_{k+j}\epsilon_{t-j} \qquad (9.56)$$

and since

$$v_{t+k} = \sum_{j=0}^{\infty} \psi_j \epsilon_{t-j}$$

we have

$$v_{t+k} = F_k(t) + e_k(t), \qquad (9.57)$$

where $e_k(t)$ is the error made at time t in forecasting k units ahead. One could, of course, define the error in that way; the point is that this is optimal for forecasts based on linear functions of ϵ or v in a least-squares sense.

9.24　Subject to what we have said earlier about errors due to having the wrong model, it follows that the variance of the forecasting error k units ahead, from (9.57) and (9.58), is

$$\text{var } \epsilon\{1 + \psi_1^2 + \psi_2^2 + \ldots + \psi_{k-1}^2\}. \tag{9.58}$$

9.25 Equivalently we may write

$$F_k(t) = \sum_{j=1}^{\infty} \pi_j F_{k-j}(t) + \epsilon_{t+k}. \tag{9.59}$$

We then determine the forecast (1) by putting all future ϵ's equal to their zero expectation, (2) by determining past values of e from the one-step ahead error $u_{t-j} - F_1(t - j - 1)$, (3) by replacing u's of the future by their present forecasts at time t.

9.26 The Box–Jenkins model can also be adapted to take account of seasonal variation. Effectively, what is done is to take differences at lag 12 (for monthly data), i.e. to difference month-wise as well as from one month to the next, so that the model becomes

$$\alpha(B)(1 - B)^d(1 - B^{12})u_t = \beta(B)(1 - B^{12})\epsilon_t. \tag{9.60}$$

Further powers of $1 - B^{12}$ may be added, in which case further parameters have to be estimated.

9.27 An extensive comparison of different methods of adaptive forecasting was undertaken by Reid (1971). He took 113 time-series, including annual, quarterly and monthly data, mostly on macro-economic variables in the U.K., although some were from the U.S.A. The series were in some cases short, the quarterly figures, for example, having fewer than 60 terms. Many comparisons are possible and not every method was applied to every series (because, *inter alia*, some were not seasonal). Table 9.1, however, presents a general picture of the results. The criterion of "best" in this table is the minimization

Table 9.1 *Forecasting methods on 113 series (Reid, 1971)*

Method	Number of series		Percentage
	Method used on	Method best on	
Box–Jenkins	113	76	67
Brown	113	2	2
Modified Brown[*]	87	18	21
Winters	69	10	15
Harrison	47	7	15

[*] The "modified Brown" method is one in which Reid, finding serial correlation in the errors, fitted a first-order autoregressive series to those errors.

of forecast errors one step ahead. Reid also made comparisons for longer lead times. The results were similar, but Brown's method and Harrison's method improved somewhat, though not to overtake Box–Jenkins. On the other hand, there were certain types of series for which certain methods seem particularly appropriate; for instance, Harrison's method did particularly well on unemployment figures, which have a marked seasonal and irregular component.

9.28   It will be evident that the efficiency of an adaptive forecasting method depends on a number of factors, and the practitioner has a fairly wide choice, not only among types of model, but among the various degrees of parametrization incorporated in it. Minimization of errors is only one of the criteria which he will wish to employ. Others are

(1) The amount of effort involved in fitting and the ready availability of machine programs.

(2) The speed with which a method picks up a significant change in behaviour, e.g. a sudden shift in mean or a rise in the slope of the trend.

(3) The existence of serial correlations in the errors (indicating, in many cases, that the model is oversimplified).

(4) The permanence of the primary data. (Many official series have the irritating property of being retrospectively altered from time to time.)

(5) The sheer volume of the work – in some fields several thousand series arise for updating every month, and economy and speed are paramount.

(6) Urgency – in areas such as traffic control, power loads, and missile tracking the forecast has to be practically "on line", that is to say the forecast, if it is to be any use, must be obtainable in a matter of hours or minutes or even seconds.

Choice of predictive method

9.29   Reid, in the work under reference, gives a decision tree for choosing a predictive method, and although further experience may lead to a modification of its particular recommendations, especially in non-economic fields, the concept is a useful one. Perhaps every practitioner needs his own tree; Reid's is shown in Fig. 9.1.

Kalman filters

9.30   The methods we have considered up to this point were developed in the context of economics or business (Holt, Winters, Harrison) or in engineering control theory (Box and Jenkins). For some years, engineers, following a basic paper of Kalman (see, for example, Kalman and Bucy, 1961), pursued

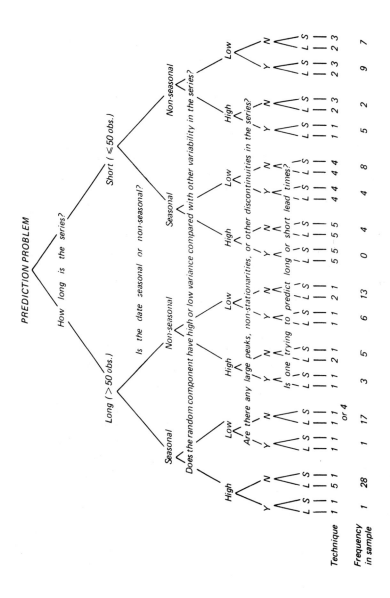

Fig. 9.1   A method of choosing a time-series prediction technique.

Key to techniques: 1: Box–Jenkins; 2: Modified Brown; 3: Brown; 4: Holt–Winters; 5: Harrison

an independent course by developing recursive methods for multisensor systems, for example in tracking mechanisms or servo-controls. Effectively the problems to be considered, from many points of view, are the same as those of forecasting by more familiar methods, but there are some important differences:

(a) In the engineering context there is often a great deal of prior information which enables close approximations to be made to the parameters of the system; or alternatively a great deal of information about other systems of the same kind, which serves the same purpose.

(b) The input of information is usually multidimensional.

(c) Sometimes the actual "prediction" cannot be verified against observation. For example, in tracking a moving object from some stationary observation point, the incoming signals may be angles, radar reflections, and so forth, the object being to predict the position and velocity of the object, which may never be directly observed.

9,31   The utility of these methods outside the realm of control engineering has yet to be demonstrated. A beginning has been made in a paper by Harrison and Stevens (1971). Their model, in generality, considers an $m \times 1$ vector y_t, dependent through a known $m \times p$ matrix V on p parameters:

$$y_t = V\theta_t + \varepsilon_t, \tag{9.61}$$

where the θ's are supposed to obey a relation of Markovian character through an $m \times m$ known matrix L, a known $m \times p$ matrix M and residuals Y:

$$\theta_t = L\theta_{t-1} + MY_t. \tag{9.62}$$

It is assumed that at time t we know V, L, M and the probability distribution of ε_t and Y_t. It is also assumed that ε, Y and θ are multinormal. The argument then proceeds by finding the posterior distribution of θ_t based on the prior distribution θ_{t-1} and the additional information accruing from time $t-1$ to time t; or rather, the mean and dispersion of θ_t given the estimators at $t-1$ and the errors. For complicated, or even perhaps for relatively simple systems, the amount of prior knowledge which has to be obtained or assumed is obviously substantial.

NOTE

A readable account of Kalman methods appears in the *Course* (1971) given under TASC in the references. See also Harrison and Stevens (1971).

10

Multivariate series

10.1   Suppose that we have several series moving concomitantly through time and wish to consider the relationships among them. We may represent them by a $p \times n$ matrix, i.e. p series observed at n points of time. To try to decompose them into trend, seasonality, etc., by a simultaneously performed analysis makes for almost unmanageable arithmetic. We will suppose that each of the p components has been considered separately in regard to trend and seasonal movements and that these have been abstracted; or alternatively that trend has been dealt with by differencing and seasonality abstracted. We are then left to consider the relationships among stationary series, and in particular the relationship between a pair of series. (With more series we shall usually have to consider them in pairs.)

Cross-correlation

10.2   We can now generalize to a pair of series the notion of autocorrelation, which we shall call *cross-correlation*. Some authors, we may remark in passing, refer to such relations as "serial correlations", a term which we have reserved for the observed correlations of a particular finite realization. The covariance of u_{it} $(i = 1, 2, \ldots, p)$ and $u_{j, t-s}$ will be denoted by $\gamma_{(ij)s}$. Where a pair $1, 2$ only are concerned we may omit the suffixes (i, j). The corresponding parental cross-correlation will be denoted by $\rho_{(ij)s}$. The observed values, as usual denoted by Roman letters, are $c_{(ij)s}$ and $r_{(ij)s}$. For any given s there are $\frac{1}{2}p(p + 1)$ of these quantities which we denote by the square matrix Γ_s, P_s, c_s or r_s as the case may be.

10.3   In the univariate case, as we have seen, $\rho_k = \rho_{-k}$, so that only positive values of k need to be considered. For two or more series this is no longer so. Clearly the correlation between u_1 and u_2 when the former leads the latter by s units is not the same as when it lags behind by s units. Thus Γ_s is not symmetric about its diagonal. On the other hand, since $E(u_{1,t} u_{2,t+s})$ is equal to $E(u_{1,t-s} u_{2,t})$, we have

$$\Gamma_s = \Gamma'_{-s}. \tag{10.1}$$

129

Cross-spectra

10.4   We may define a cross-spectrum between two series as

$$w_{12}(\alpha) = \sum_{s=-\infty}^{\infty} \rho_{(12)s}\, e^{i\alpha s}, \tag{10.2}$$

with a corresponding integrated spectral function $W(\alpha)$ defined over the range 0 to π. Conversely we have for the cross-correlations

$$\rho_{(12)s} = \frac{1}{2\pi} \int_{-\pi}^{\pi} w_{12}(\alpha)\, e^{-i\alpha s}\, d\alpha. \tag{10.3}$$

In univariate formulae, as we have seen, the sine terms in these expressions cancel in virtue of the fact that $\rho_k = \rho_{-k}$ and the spectral density is real. This is now no longer the case. From (10.2) we have

$$w_{12}(\alpha) = \rho_{12}(0) + \sum_{1}^{\infty} \{\rho_{(12)s} \cos s\alpha + \rho_{(12)-s} \cos s\alpha\}$$

$$+ i\left\{\sum_{1}^{\infty} \rho_{(12)s} \sin s\alpha - \rho_{(12)-s} \sin s\alpha\right\} \tag{10.4}$$

$$= c(\alpha) + iq(\alpha), \tag{10.5}$$

where

$$c(\alpha) = 1 + \sum_{s=1}^{\infty} \cos s\alpha\{\rho_{(12)s} + \rho_{(12)-s}\} \tag{10.6}$$

$$q(\alpha) = \sum_{s=1}^{\infty} \sin s\alpha\{\rho_{(12)s} - \rho_{(12)-s}\}. \tag{10.7}$$

Coherence

10.5   The spectral density now has an imaginary as well as a real component The quantity $c(\alpha)$ is usually called the *co-spectrum* or *co-spectral density*. $q(\alpha)$ is called the *quadrature spectrum* or *spectral density*. The sum of squares $c^2(\alpha) + q^2(\alpha)$ is called the *amplitude* of the spectrum. Standardized by division by the separate spectral densities of the two series it is called the *coherence*, namely

$$C(\alpha) = \frac{c^2(\alpha) + q^2(\alpha)}{w_1(\alpha) w_2(\alpha)}. \tag{10.8}$$

The phase relationship of these quantities also requires attention. The so-called phase diagram is the graph of $\psi(\alpha)$ as ordinate against α as abscissa, where

$$\psi(\alpha) = \arctan \frac{q(\alpha)}{c(\alpha)}. \tag{10.9}$$

The Argand diagram plots $c(\alpha)/w_1(\alpha)$ as abscissa against $q(\alpha)/w_1(\alpha)$ as ordinate. The gain diagram plots α as abscissa against an ordinate $R_{12}^2(\alpha)$, where

$$R_{12}^2(\alpha) = \frac{w_1(\alpha)}{w_2(\alpha)} C(\alpha). \tag{10.10}$$

The coherence $C(\alpha)$ can be thought of as analogous to a correlation coefficient, the gain $R^2(\alpha)$ as analogous to a regression coefficient.

10.6   The interpretation of these quantities in a time-series context, so it seems to me, is more difficult than that of the cross-correlogram, which at least displays the intensity of relationship between two series for various leads or lags. Broadly speaking, the coherence purports to measure the degree to which the two series vary together, and the phase the extent to which they are in step. For a fuller treatment of the subject reference may be made to the book by Granger and Hatanaka (1964). A recent attempt to apply the technique to import–export statistics is made by Gudmundsson (1971).

Example 10.1

Table 10.1, for comparison with Table 1.6, gives the U.K. production of motor vehicles for each quarter of 1960-1971. There are no seasonal elements in these two series. In Fig. 10.1 we show the two series (which happen to have about the same range in the ordinate) after removal of the irregular component by a Census X–11 program. The series as graphed then comprise only trend and short-term oscillatory movements. A glance at the diagram suggests that, although the two series pursue far from identical courses, there is some similarity in the time of occurrence of peaks and troughs.

There are various ways in which we might proceed to eliminate trend, which is not very marked in either case. For simplicity in this example a linear trend was removed from both series. Fig. 10.2 shows the correlograms of the two resultant series. In both cases the oscillatory effects are marked. The cross-correlogram is shown in Fig. 10.3. Here, at zero lag, the correlation is relatively small but positive, being about 0·33. At F.T. Index leading the car production by about 7 quarters it is negative and about −0·45. Again at the car production leading F.T. Index by about 7 quarters it has a minimum of about −0·22.

One can provisionally interpret this in two ways. Either one series leads the other by about 16 quarters (4 years) or both have, independently, a rhythm of about that period.

In Fig. 10.4 we have re-graphed the original series but have reversed the sign of one of them (the F.T. Index) and led it seven quarters in front of car production. The coincidence of peaks and troughs is now brought out more sharply. The indication is that a peak in the F.T. Index is followed seven quarters later (or thereabouts) by a trough in car production.

The interesting question, of course, is what is implied by this result. It would be premature to conclude that a peak in the F.T. Index *causes* a slump in car production about two years later. We may have an economy with an oscillation in each series averaging about four years in duration, the oscillations being to some extent in step (as measured by the correlation at zero lag). The

Table 10.1 *U.K. Production of Cars, seasonally adjusted (thousands)*

Year and Quarter		Car Production	Year and Quarter		Car Production	Year and Quarter		Car Production
1960	1	374·241	1964	1	460·397	1968	1	435·000
	2	375·764		2	462·279		2	435·000
	3	354·411		3	434·255		3	465·000
	4	249·527		4	475·890		4	483·000
1961	1	206·165	1965	1	439·365	1969	1	423·000
	2	258·410		2	431·666		2	427·990
	3	279·342		3	399·160		3	425·490
	4	264·824		4	449·564		4	438·230
1962	1	312·983	1966	1	435·000	1970	1	407·000
	2	300·932		2	435·000		2	416·000
	3	323·424		3	408·000		3	332·000
	4	312·780		4	330·000		4	469·320
1963	1	363·336	1967	1	369·000	1971	1	403·750
	2	378·275		2	390·000		2	416·046
	3	414·457		3	381·000		3	435·236
	4	459·158		4	414·000		4	455·897[(*)]

[(*)]First two months expressed at a quarterly rate.

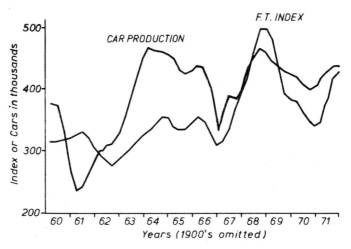

Fig. 10.1   F.T. Index and car production in U.K. (both smoothed by removal of irregular component)

analysis is suggestive of search for causal connection, and relations of this kind, if stable, can be used for forecasting, as we shall see in the next chapter, even when the causal linkage is not overt. In any case, we have here a series of only 48 terms extending over 12 years. Nor do ordinary tests of significance of individual correlations help very much, because we have deliberately picked out the biggest for attention.

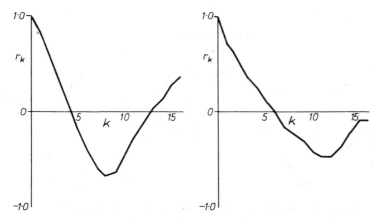

Fig. 10.2   Correlograms of F.T. Index (*left*) and car production (*right*): original series with irregular component and linear trend removed

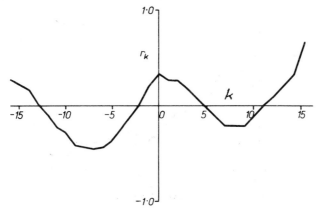

Fig. 10.3   Cross-correlogram, F.T. Index *v.* car production: original series, irregular component and linear trend removed

Let us consider the relationship between the series from the point of view of spectrum analysis. In Fig. 10.5 are given the spectra of the two series, and in Fig. 10.6 their coherence. The individual spectra of Fig. 10.5 are not very informative. There is no evidence of "cycles" in any strict sense. The higher values on the left indicate mean periods of about 16 quarters for the F.T. Index (a cycle-frequency of about 0·6) and rather longer for car production, which confirms the picture presented by the correlograms. In the coherence diagram two peaks are suggestive, one at cycle-frequency about 0·09, the other at about 0·27, corresponding to mean periods of about 11 and 4 quarters respectively. The interpretation of these results is more difficult. It must always be remembered that they are sample values and therefore liable to

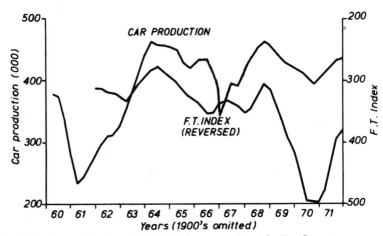

Fig. 10.4   Car production, with F.T. Index reversed and leading 7 quarters

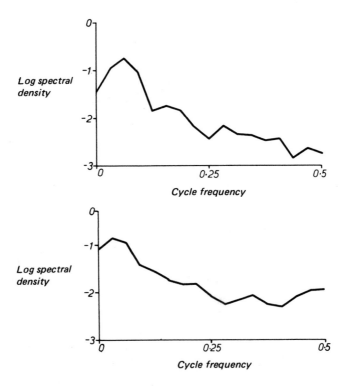

Fig. 10.5   Spectrum of F.T. Index (*upper*) and car production (*lower*): original series, irregular component and linear trend removed; Bartlett smoothing

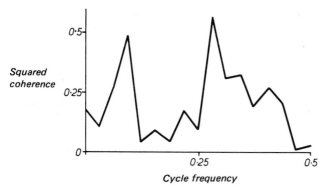

Fig. 10.6   Coherence of F.T. Index and car production: original series, irregular component and linear trend removed

considerable fluctuation. The one-yearly component seems to imply some common movement within the year, notwithstanding that one series (car production) is supposed to be seasonally adjusted. The other is presumably a reflection of the fact that both series have an oscillatory element of several years, and these elements, within the time-period of 12 years, are to some extent in step.

In series generated by physical processes, especially where fundamental generators of a harmonic kind are operating, the interpretation of spectra is usually more straightforward. In economic series there are considerable difficulties, especially where the series are short. The practitioner who uses methods based on cross-correlograms or cross-spectra needs a good deal of experience before he can draw firm conclusions from them.

Types of models

10.7   As in the univariate case, we shall be concerned with three types of model, autoregressive, moving-average, and mixed. A little care is required in the formulation. For example, with autoregression we now have p series, and for any one of these the value at time t is related not only to its own values at previous times, but to those of the other series; and for each of the p relationships there is a residual element ϵ, which may differ from one series to another. For a series of order k we then have $p + kp^2$ constants α, even if the ϵ's are independent and not autocorrelated, and are clearly heading for an embarrassing profusion of constants if p and k are large.

Consider, for example, three Markoff type series u_{1t}, u_{2t}, u_{3t}. We shall now keep the coefficients of the first terms u_{it}, not equate them to unity. In full, then, the schemes may be written

$$\alpha_{01}u_{1t} + \alpha_{11}u_{1,t-1} + \alpha_{21}u_{2,t-1} + \alpha_{31}u_{3,t-1} = \epsilon_{1t} \qquad (10.11)$$

$$\alpha_{02}u_{2t} + \alpha_{12}u_{1,t-1} + \alpha_{22}u_{2,t-1} + \alpha_{32}u_{3,t-1} = \epsilon_{2t} \qquad (10.12)$$

$$\alpha_{03}u_{3t} + \alpha_{13}u_{1,t-1} + \alpha_{23}u_{2,t-1} + \alpha_{33}u_{3,t-1} = \epsilon_{3t}, \qquad (10.13)$$

or, with the backward shift operator,

$$(\alpha_{01} + \alpha_{11}B)u_{1t} + \alpha_{21}Bu_{2t} + \alpha_{31}Bu_{3t} = \epsilon_{1t} \qquad (10.14)$$

$$\alpha_{12}Bu_{1t} + (\alpha_{02} + \alpha_{22}B)u_{2t} + \alpha_{32}Bu_{3t} = \epsilon_{2t} \qquad (10.15)$$

$$\alpha_{13}Bu_{1t} + \alpha_{23}Bu_{3t} + (\alpha_{03} + \alpha_{33}B)u_{3t} = \epsilon_{3t}, \qquad (10.16)$$

or, equivalently in matrix form,

$$(\mathbf{A}_0 + \mathbf{A}_1 \mathbf{B})\mathbf{u}_t = \boldsymbol{\epsilon}_t, \qquad (10.17)$$

where

$$\mathbf{A}_0 = \begin{pmatrix} \alpha_{01} & 0 & 0 \\ 0 & \alpha_{02} & 0 \\ 0 & 0 & \alpha_{03} \end{pmatrix}, \quad \mathbf{A}_1 = \begin{pmatrix} \alpha_{11} & \alpha_{21} & \alpha_{31} \\ \alpha_{12} & \alpha_{22} & \alpha_{32} \\ \alpha_{13} & \alpha_{23} & \alpha_{33} \end{pmatrix}, \qquad (10.18)$$

$\mathbf{u}_t$ is the column vector u_{1t}, u_{2t}, u_{3t}, and $\boldsymbol{\epsilon}_t$ is the row vector $\epsilon_{1t}, \epsilon_{2t}, \epsilon_{3t}$.

10.8 We shall therefore write the general linear autoregressive scheme as

$$A(B)\mathbf{u}_t = \boldsymbol{\epsilon}_t, \qquad (10.19)$$

where

$$A(B) = \sum_{s=0}^{k} \mathbf{A}_s B^s \qquad (10.20)$$

and the $\mathbf{A}$'s are $p \times p$ matrices of coefficients.

Formally we may write the solution of this difference equation (systematic terms being supposed to have damped out) as in the univariate case at equation (6.41), as

$$\mathbf{u}_t = \mathbf{A}^{-1}(B)\boldsymbol{\epsilon}_t. \qquad (10.21)$$

The inverse $\mathbf{A}^{-1}$ will in general consist of the quotient of two polynomials in B. The denominator expanded and multiplied by the numerator gives us an infinite series of constants acting as weights in the infinite average of the ϵ's.

10.9 Likewise we may write the moving-average process in terms of matrices of coefficients, denoted by $\mathbf{M}$. Thus,

$$\mathbf{u}_t = \sum_{s=0}^{l} \mathbf{M}_s B^s \boldsymbol{\epsilon}_t = \mathbf{M}(B)\boldsymbol{\epsilon}_t. \qquad (10.22)$$

The mixed process is obviously

$$A(B)u_t = \mathbf{M}(B)\boldsymbol{\epsilon}_t. \qquad (10.23)$$

Now we suppose that the ϵ's all are standardized to zero mean and variance σ^2. Since we have left the coefficients α_{0i}, β_{0i} undetermined, no generality is lost by this re-scaling. We also suppose the ϵ's independent. Then for the dispersion matrix we have

$$\gamma_s = \mathrm{E}(\mathbf{u}_t \mathbf{u}'_{i-s})$$

and substituting from (10.22) for the moving-average scheme,

$$\gamma_s = E\left\{\sum_{j=0}^l \mathbf{M}_j B^j \boldsymbol{\epsilon}_t\right\}\left\{\sum_{m=0}^l \mathbf{M}_m B^m \boldsymbol{\epsilon}_t\right\}$$

$$= \sum_{j=0}^l \sum_{m=0}^l \mathbf{M}_j \mathbf{M}'_m \, E\left(\boldsymbol{\epsilon}_{t-j}\boldsymbol{\epsilon}'_{t-s-m}\right)$$

$$= \sigma^2 \sum_{s,m} \mathbf{M}_{s+m}\mathbf{M}'_m$$

$$= \text{coeff. of } B^s \text{ in} \quad \sigma^2(\sum \mathbf{M}_j B^j)(\sum \mathbf{M}'_j B^{-j}). \tag{10.24}$$

This is the generalization of the result at equation (6.66). Likewise for the autoregressive scheme,

$$\gamma_s = \text{coeff. of } B^s \text{ in} \quad \mathbf{A}^{-1}(B)\{\mathbf{A}^{-1}(B^{-1})\}' \tag{10.25}$$

and for the mixed scheme of (10.23)

$$\gamma_s = \text{coeff. of } B^s \text{ in} \quad \mathbf{A}^{-1}(B)\mathbf{M}(B)\mathbf{M}'(B^{-1})\{\mathbf{A}^{-1}(B^{-1})\}' \tag{10.26}$$

The relative elegance with which these relations can be written down will deceive no-one into thinking that the resulting equations are easy to solve. In fact, as we shall see later, they may not possess a unique solution.

10.10 It is natural to inquire whether the autoregressive scheme is amenable to the treatment through equations of the Yule–Walker type, which enabled us to determine, in the univariate case, the constants of the scheme in terms of the autocorrelations. We can indeed derive such equations. Postmultiplying (10.19) by $\mathbf{u}'_{t+q}$ and taking expectations, we have

$$E \sum_{s=0}^k \mathbf{A}_s B^s \mathbf{u}_t \mathbf{u}'_{t+q} = \sum \mathbf{A}_s E\left(\mathbf{u}_{t-s}\mathbf{u}'_{t+q}\right)$$

$$= \sum_{s=0}^k \mathbf{A}_s \gamma_{q+s} = 0, \quad q > 0. \tag{10.27}$$

The solution of these equations is feasible but not easy.

The unidentifiability problem

10.11 For the mixed scheme of (10.23), however, a different point arises. We noticed briefly in **6.27** that the autocorrelations of a moving-average process do not uniquely determine its constants. We may also recall the remark in **9.23** that a condition on the constants in a Box–Jenkins model is also required to achieve determinacy. The same kind of situation arises here.

Example 10.2 (Quenouille, 1957)

If we write $\mathbf{F}$ for the matrix $\mathbf{A}^{-1}\mathbf{M}$, we can then write (10.26) as

$$\gamma_s = \text{coeff. of } B^s \text{ in} \quad \mathbf{F}(B)\mathbf{F}'(B^{-1}).$$

Consider now the matrices

$$\mathbf{F}_1 = \begin{pmatrix} 2+B & B \\ 1 & 6+B \end{pmatrix}, \qquad \mathbf{F}_2 = \begin{pmatrix} 1+2B & -B \\ 5 & 3+2B \end{pmatrix},$$

$$\mathbf{F}_3 = \frac{1}{5}\begin{pmatrix} 2+7B & 4+9B \\ -11-6B & 28+3B \end{pmatrix}, \qquad \mathbf{F}_4 = \frac{1}{17}\begin{pmatrix} 14+37B & -5-12B \\ 55+30B & 1+84B \end{pmatrix}. \quad (10.28)$$

It can easily be verified that

$$|F_1| = (4+B)(3+B), \qquad |F_2| = (3+B)(1+4B)$$
$$|F_3| = (4+B)(1+3B), \qquad |F_4| = \tfrac{1}{17}(1+4B)(1+3B) \quad (10.29)$$

and that for each $\mathbf{F}$, $\mathbf{FF}'$ is

$$\begin{pmatrix} 6+2B+2B^{-1} & 3+7B \\ 3+7B^{-1} & 38+6B+6B^{-1} \end{pmatrix}. \quad (10.30)$$

Thus several schemes may give rise to the same dispersion matrix and, in the language of econometrics (where a similar problem arises), the situation is "unidentifiable". It does not follow that all the solutions are acceptable; some of them might lead to non-stationary series, but it seems that some fairly restrictive conditions must be imposed to ensure that the solution is unique. A general discussion is given by Phillips (1959) and Kendall and Stuart (vol. 3) but is complicated and not very helpful in practice.

10.12   A complete and practically convenient solution to the unidentifiability problem is still awaited. The root of the problem, oddly enough, lies in the moving-average, not the autoregressive part of the model. In these circumstances it is worth while considering whether we can, by approximation, get rid of the moving-average component.

Consider, for example, an autoregressive scheme in which the residuals are themselves autoregressive, e.g. the univariate case

$$\sum_{j=0}^{k} \alpha_j u_{t-j} = \eta_t \quad (10.31)$$

where

$$\sum_{m=0}^{l} \beta_m \eta_{t-m} = \epsilon_t, \quad (10.32)$$

ϵ_t being a random variable. We can substitute for η_t from (10.31) in (10.32) to obtain

$$\sum_{m=0}^{l} \beta_m \sum_{j=0}^{k} \alpha_j u_{t-m-j} = \epsilon_t, \quad (10.33)$$

and the series is now purely autoregressive. Its constants can be estimated in the usual way. They will, however, be mixed functions of α and β and some care may be required to disentangle the two. It may, in fact, be preferable to regard the residual η as autoregressive rather than consider its correlations as generated by a moving average. Or, to put it another way, a more extended purely autoregressive scheme of extent $k + l$ (without bothering to disentangle α and β) may be preferable to the mixed scheme with moving-average residuals. These considerations apply *a fortiori* to multiple series. We revert to the topic in Chapter 12.

NOTE

A report from the Institute for Mathematical Studies in the Social Sciences of Stanford University by F.C. Nold (1972) gives a bibliography of applications of spectral analysis to economic time-series.

11

Forecasting from lagged relationships

11.1 In many situations wherein we have several concomitant variables linked by some causal mechanism there is reason to suppose that the causal variables lead (i.e. happen before) the effects. We need not worry unduly about the strange philosophical problems connecting cause and effect. In practice, things happen to generate other events later in time, albeit passing by a roundabout chain of causality through a transient phase. If a variable in which we are interested can be related to other leading variables in a systematic way, we obviously have the material for a forecasting method. Sometimes our knowledge of the situation tells us which are the leading variables. More often we have to search for them and for the time-extent of the lead. Sometimes there are considerations both ways: for example, a rise in prices leads to a rise in wages, but a rise in wages will also lead to a rise in certain prices.

11.2 Where we are dealing with a highly interactive system, such as we encounter in economics, we may have to construct a model consisting of a number of equations, in some of which the variable x will lead the variable y and in others of which x will lag behind y. The theory of constructing and analysing such models lies outside the scope of this book. We shall concentrate on single-equation models. Given a variable of interest, y at time t, we shall try to set up an equation in terms of other variables x_1, x_2, etc., observed at previous times. (Some of the x's may be lagged values of y.) The use of such an equation for forecasting is then obvious.

11.3 Suppose that we have a system consisting of interacting variables u (known as endogenous) and variables v (exogenous) which influence the system but are not influenced by it. Suppose further that the variables are linked by a set of linear equations which may, *inter alia*, include certain lagged values of the u's. The system can be expressed in matrix form as

$$Au + Hv = \varepsilon,$$ (11.1)

where the coefficients in **A** may be polynomials in the shift operator B; and so for **H**. The determinant of **A** will, in general, be a polynomial in B and the inverse of **A** will be the quotient of two polynomials, **P/Q** say. Then

$$u = -\mathbf{A}^{-1}\mathbf{H}v + \mathbf{A}^{-1}\epsilon,$$

and on multiplying by Q we see that Qu, an autoregressive expression in u, is equal to some linear function of the v's, including perhaps lagged terms, plus some linear function of the ϵ's. We may therefore expect that any component of u may be expressed in terms of lagged values of u, together with (perhaps) contemporary and lagged values of the v's plus a stochastic residual. It is an equation of this kind that we seek, except that, for forecasting purposes, we want all the variables, other than u_t itself, to be lagged by at least one unit.

11.4   The problems are fourfold: (1) to determine which are the right variables; (2) to determine at what lags they should enter the equation; (3) to set up the equation in some optimal way; (4) to validate the result, i.e. to justify its use for the future. We might perhaps add (5): to explain why the equation works (i.e. acts as a successful prediction) — if it does. So long as we can rely on its permanence this is, perhaps, not strictly necessary, but is always desirable. And indeed, so long as the equation remains a purely empirical one, our confidence in its permanence can hardly be very high, especially in times of economic change.

Discarding variables

11.5   In general, we can write down many variables x which, on *a priori* grounds, are thought to have some possible influence on y, and we can write them at various lags in time. If we put them all, at a number of lags, into a regression equation, we shall probably find many of them redundant in the sense that they add little or nothing to the fit of the regression as measured by the size of the multiple correlation coefficient or, equivalently, the smallness of the residual variance. The discarding of variables has to be carried out with some care, but it is common experience, at least in economic work, that a great many can be discarded without serious harm to the adequacy of the regression equation. This procedure does not necessarily imply that the discarded variables are unimportant; only that they are so highly correlated with the retained variables that the latter are, so to speak, acting on their behalf.

Stepwise forward and stepwise backward procedures

11.6   Two routines in common use for the discarding of redundant variables in regression analysis are known as "stepwise forward" and "stepwise backward" procedures. In the former we regress y on each of the variables $x_1, x_2,$ etc., separately. One of these p regressions will be better than the others (in the sense that it gives the smallest sum of squares of residuals) and the corresponding variable, say x_1, is then brought in and retained. At the next stage, y is regressed on x_1 and, in turn, each of the other $p - 1$ variables. Again one of these will be better than the others, say that involving x_2, which is then retained in the analysis. The third stage considers the regression of y on x_1, x_2 and, in turn, each of the remaining $p - 2$ variables; and so on, step by step, until enough variables have been brought in to give a satisfactory fit, as measured by R^2. The other variables are discarded.

Stepwise backward methods proceed similarly, except that they begin by regressing y on all the p variables and then seek for the one which, on being discarded, affects the regression least in the sense of lowering R^2 as little as possible. Then the surviving $(p - 1)$ variables are examined to see which can be discarded, and so on, again until a satisfactory value of R^2 remains but no further variable can be discarded without serious loss.

The optimal regression method

11.7 Unfortunately these methods do not always work satisfactorily. They may not give the same answer, and even if they do the answer may be far short of optimal; that is to say, if the two methods indicate that a certain set of k variables give an acceptable value of R^2 there may be some other set or sets of k which give a considerably higher value. A preferable method is to employ a routine which picks out the optimal set, i.e. a routine that, for each value of k from 1 to p, will determine which set of variables, regressed with y, has the highest value among all sets of k. For such a method reference may be made to Beale, Kendall and Mann (1967). We shall employ it in one of the examples below.

Some practical examples

Example 11.1

Table 11.1 gives the value of U.K. imports (I) for each quarter of the years 1960—70 corrected for seasonality, together with other variables which might be expected to influence imports, namely stocks (S), fixed investments (F), and consumer durable expenditure (D). The question is whether we can use the information available up to and including one quarter to predict the imports for the following quarter. It seems unlikely that the other variables will influence imports more than one year ahead, so we shall consider the regression of I_t on $I_{t-1}, I_{t-2}, I_{t-3}, I_{t-4}$ and the similarly lagged values of S, F and D.

When fitting as many variables as this to relatively short series, we have to pay some attention to the number of degrees of freedom involved in the estimates, especially those of R^2. For example, if we fit a regression to n members with p variables the residual is usually estimated as

$$\text{est var } \epsilon = \frac{S_R}{n - p}, \tag{11.2}$$

Table 11.1   *U.K. imports for each quarter of the years 1960–1970*

Year	Quarter	I	S	D	F	
1960	1	1382	149	370	1088	I = Imports of goods
	2	1417	168	342	1081	and services
	3	1432	161	332	1103	
	4	1438	150	307	1146	S = Value of physical
1961	1	1457	153	327	1184	increases in stocks
	2	1403	102	331	1205	and work in pro-
	3	1389	35	329	1241	progress
	4	1379	37	313	1217	
1962	1	1408	7	316	1197	D = Consumer expen-
	2	1426	12	361	1221	diture on durables
	3	1460	53	336	1222	
	4	1442	−6	350	1189	F = Gross domestic
1963	1	1414	−3	353	1070	fixed capital
	2	1472	36	398	1247	formation
	3	1520	−1	414	1278	
	4	1540	159	416	1317	
1964	1	1611	132	417	1373	
	2	1612	170	429	1417	
	3	1632	141	439	1459	
	4	1659	185	440	1476	
1965	1	1581	92	453	1486	
	2	1643	83	426	1474	
	3	1672	112	428	1471	
	4	1686	89	417	1529	
1966	1	1722	86	447	1500	
	2	1681	76	468	1509	
	3	1726	86	409	1551	
	4	1642	6	375	1552	
1967	1	1777	60	392	1587	
	2	1787	64	428	1671	
	3	1779	−1	467	1645	
	4	1850	61	492	1621	
1968	1	1948	−104	550	1721	
	2	1903	76	408	1693	
	3	1945	85	434	1714	
	4	1937	101	460	1722	
1969	1	1992	122	400	1710	
	2	1980	76	425	1672	
	3	1966	57	444	1705	
	4	2024	91	432	1690	
1970	1	2026	−16	433	1646	
	2	2130	116	457	1759	
	3	2078	117	476	1726	
	4	2197	112	480	1755	

All data are seasonally adjusted. Data from *Monthly Digest of Statistics*. Figures in £ million.

where S_R is the residual sum of squares. If S_T is the total sum of squares, the variance of y, the regressand, is estimated as S_T/n, and

$$\text{est } R^2 = 1 - \frac{S_R}{n-p} \frac{n}{S_T}. \tag{11.3}$$

If a mean-term is to be included, n is reduced by unity. The estimators $S_R/(n-p)$ and S_T/n are then unbiassed, and the estimator of R^2 in (11.3) will be virtually so. However, the addition of a variable to an equation (which must lower the residual sum of squares because an extra constant is fitted) may actually increase the estimated R^2. Indeed, if the regression is a very poor fit, equation (11.3) may even give a negative value.

It is therefore also useful to compute the maximum likelihood estimator

$$\mathrm{ML}\,(R^2) = 1 - \frac{S_R}{S_T} \qquad (11.4)$$

which cannot be negative, but may be biassed. In the following we shall tabulate both forms.

Let us first consider the regression of I_t on the four lagged terms of I itself, namely an autoregression, without regard to the other variables. Since we have a lag of four quarters to consider, the first value of I_t must be that for the first quarter of 1961. There will then be 40 data points, i.e. the 10 years 1961 through 1970. The regression is

$$I_t = -51{\cdot}7799 + 0{\cdot}503\,74 I_{t-1} + 0{\cdot}469\,83 I_{t-2} + 0{\cdot}020\,48 I_{t-3} + 0{\cdot}054\,23 I_{t-4}. \qquad (11.5)$$

est R^2 is 0·9607 and ML (R^2) is 0·9647.

The contribution of I_{t-3} and I_{t-4} is small, and a fair predictor could be based on I_{t-1} and I_{t-2} alone. Fig. 11.1 shows the actual series and the values "predicted" by equation (11.5).

It is of some interest to consider what happens if we fit I_t to a subset I_{t-1} to I_{t-4} by the optimal regression program. The results are as follows:

Constant	I_{t-1}	I_{t-2}	I_{t-3}	I_{t-4}	est R^2	ML (R^2)	
−13·1650	1·01898	—	—	—	0·9529	0·9762	(11.6)
−44·4579	0·52751	0·51485	—	—	·9627	·9646	(11.7)
−48·5274	0·50927	·47549	—	0·06299	·9618	·9647	(11.8)
−51·7799	0·50374	·46983	0·02048	·05423	·9607	·9647	(11.9)

The increment to R^2 on adding I_{t-4} and I_{t-3} is so slight (and indeed is a decrement for est R^2) that we are justified in concluding that the autoregressive equation specified by (11.7) is adequate, and even (11.6) is quite good.

It should be noticed that neither (11.6) nor (11.7) gives us a *stationary* series. Equation (11.6), considered as a Markoff type scheme, has a parameter ρ greater than unity. Likewise it will be found that one of the roots of the equation, derived from (11.7),

$$x^2 - 0{\cdot}527\,51x - 0{\cdot}514\,85 = 0, \qquad (11.10)$$

is also greater than unity; and this series also is not stationary. A glance at

Fig. 11.1 will confirm these conclusions.

Measured by R^2, the equation of (11.7) indicates that the residual variance is less than 4% of the variance of the original series. However, this is a somewhat deceptive criterion of the excellence of the fit because the whole series is evolutionary and has a large variance. We are therefore led to consider whether we should do better by working with the first differences of the series of I, and hence using the regression equation to predict the *increment* of I from one quarter to the next. The results are exhibited below. We now have 41 observations. The regressand variable in all cases is I_t.

Constant	ΔI_{t-1}	ΔI_{t-2}	ΔI_{t-3}	ΔI_{t-4}	est R^2	ML (R^2)	
26·7333	−0·47281	−	−	−	0·1788	0·2004	(11.11)
27·5775	− ·47452	−	−	−0·04958	·1580	·2023	(11.12)
26·8944	− ·46280	0·02997	−	− ·05234	·1346	·2030	(11.13)
27·2320	− ·46321	·02596	−0·01233	− ·05639	·1093	·2030	(11.14)

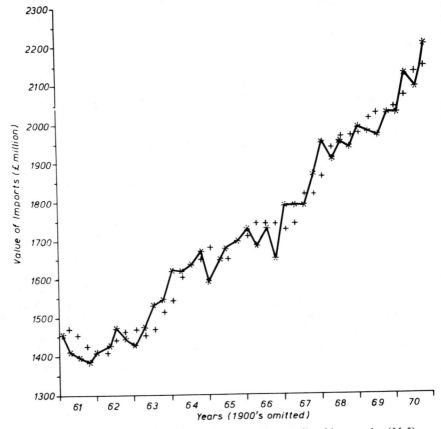

Fig. 11.1   Graph of U.K. imports (Table 11.1) and values predicted by equation (11.5)

Actual = *   Predicted = +

As with the original series, the optimal regression brings in terms in ΔI_{t-4} before those of ΔI_{t-3}, but the effect of both is negligible. Indeed the values of est R^2 diminish as we add more terms, an effect which, we noted above, may happen with short series. We should conclude that the first differences (so far as linear autoregression goes) are represented by a Markoff type process with parameter -0.4728, and that this accounts for only about 20 per cent. of the variation. The predictors from first differences show therefore little improvement on those for the original series. This is confirmed by a scrutiny of the actual residuals (which we omit to save space). The four-term equation specified by (11.5) or (11.9), over the 40 values 1961 through 1970, had a root-mean-square residual of about 43. Over the 39 values of first differences the root-mean-square residual derived from (11.14) was about 45.[*]

Let us now consider whether we can improve the regression by bringing into play the other variables. We first of all work with the original series by stepwise backward elimination of the variables. We number the variables as follows:

$$
\begin{array}{llll}
I_{t-1}\ 1 & S_{t-1}\ 5 & D_{t-1}\ 9 & F_{t-1}\ 13 \\
I_{t-2}\ 2 & S_{t-2}\ 6 & D_{t-2}\ 10 & F_{t-2}\ 14 \\
I_{t-3}\ 3 & S_{t-3}\ 7 & D_{t-3}\ 11 & F_{t-3}\ 15 \\
I_{t-4}\ 4 & S_{t-4}\ 8 & D_{t-4}\ 12 & F_{t-4}\ 16
\end{array}
$$

Variables removed (out of 16)	est R^2	ML (R^2)
None	0·9589	0·9757
8	·9605	·9757
8, 14	·9618	·9755
8, 13, 14	·9631	·9754
3, 8, 13, 14	·9641	·9751
3, 8, 9, 13, 14	·9651	·9750
3, 6, 8, 9, 13, 14	·9659	·9746
3, 4, 6, 8, 9, 13, 14	·9647	·9728
3, 4, 6, 7, 8, 9, 13, 14	·9643	·9716
3, 4, 6, 7, 8, 9, 10, 13, 14	·9639	·9703
3, 4, 6, 7, 8, 9, 10, 11, 13, 14	·9638	·9694
3, 4, 6, 7, 8, 9, 10, 11, 12, 13, 14	·9637	·9683
3, 4, 6, 7, 8, 9, 10, 11, 12, 13, 14, 15	·9630	·9668
3, 4, 6, 7, 8, 9, 10, 11, 12, 13, 14, 15, 16	·9628	·9665
3, 4, 5, 6, 7, 8, 9, 10, 11, 12, 13, 14, 15, 16	·9627	·9646

[*]For what it is worth, the Durbin–Watson d-statistic (see below, **12.13**) for the equation of (11.9) was 2·0114. But this does not imply that the residuals in the original series were effectively random because the statistic does not apply to autoregressive schemes without modification.

The routine has, in fact, rejected all but I_{t-1} and I_{t-2} as not significantly contributing to the regression. The run of the values of R^2 confirms that no improvement is obtained by bringing in the other variables. This is, perhaps, a little surprising at first sight, but there are, in fact, very high correlations among some of the variables.

It is interesting to compare the results of the stepwise backward method with the optimal regression method applied to the same data. The latter, in fact, proceeds forward by finding the best single regressor, the best pair, and so on, but we will write the results in inverse order for comparison with the stepwise procedure.

Variables removed	est R^2	ML (R^2)
None	0·9589	0·9757
8	·9605	·9757
8, 14	·9618	·9755
8, 13, 14	·9631	·9754
3, 8, 13, 14	·9641	·9751
3, 8, 9, 13, 14	·9651	·9750
3, 6, 8, 9, 13, 14	·9659	·9746
3, 7, 8, 10, 11, 13, 14	·9649	·9730
3, 7, 8, 10, 11, 13, 14, 15	·9650	·9722
3, 5, 8, 10, 11, 12, 13, 14, 15	·9649	·9712
3, 5, 7, 8, 10, 11, 12, 13, 14, 15	·9652	·9706
3, 5, 7, 8, 10, 11, 12, 13, 14, 15, 16	·9653	·9697
3, 4, 5, 7, 8, 10, 11, 12, 13, 14, 15, 16	·9654	·9690
3, 4, 5, 7, 8, 9, 10, 11, 12, 13, 14, 15, 16	·9647	·9674
3, 4, 5, 6, 7, 8, 9, 10, 11, 12, 13, 14, 15, 16	·9627	·9646

So far as the first six stages of removal are concerned, the stepwise and the optimal methods agree. After that they diverge, but come together at the final stage in agreeing that the two retained variables should be I_{t-1} and I_{t-2}.

We now consider an optimal regression program applied to first differences of all the series. For simplicity we will continue to number as $1, 2, \ldots, 16$ the differences $\Delta I_{t-1}, \Delta I_{t-2}, \ldots, \Delta F_{t-4}$. The results are as follows:

Variables removed	est R^2	ML (R^2)
None	−0·0279	0·4049
14	0·0168	0·4049
3, 14	0·0577	0·4049
3, 5, 14	0·0953	0·4048
3, 5, 6, 14	0·1298	0·4046
3, 5, 6, 13, 14	0·1600	0·4031
2, 3, 5, 6, 13, 14	0·1800	0·3958
2, 3, 5, 6, 11, 13, 14	0·1913	0·3828
2, 3, 4, 5, 6, 11, 13, 14	0·2029	0·3707
2, 3, 4, 5, 6, 11, 13, 14, 15	0·2075	0·3535
2, 3, 4, 5, 6, 8, 11, 13, 14, 15	0·2175	0·3410
2, 3, 4, 5, 6, 7, 8, 11, 13, 14, 15	0·2328	0·3338
2, 3, 4, 5, 6, 7, 8, 10, 11, 13, 14, 15	0·2223	0·3042
2, 3, 4, 5, 6, 7, 8, 9, 10, 11, 13, 14, 15	0·2034	0·2663
2, 3, 4, 5, 6, 7, 8, 9, 10, 11, 12, 13, 14, 15	0·1947	0·2371
All but variable 1	0·1788	0·2004

The implication here is that no regression is very good. (We note in passing that for the whole 16 variables equation (11.3) gives a negative value of R^2 and that the bias in ML (R^2) is very high.) For what it is worth, the regression of ΔI_t on ΔI_{t-1} is

$$\Delta I_t = 26·7333 - 0·472\,81\,\Delta I_{t-1}, \qquad (11.15)$$

which is the same as (11.11). None of the other variables is effective in improving a predictor of first differences.

11.8   It will be evident that we might have done still more arithmetic on the foregoing example. It seems unlikely that we should derive any aid from the extra variables in forecasting I_t and that the two-term autoregression is as far as regression can take us. We might then, perhaps, seek for other explanatory variables; or, if we regard autoprojection as sufficient, examine the possibility of improving the forecasting power by adaptive methods.

Example 11.2

In a published study (Coen, Gomme and Kendall, 1969) an account was given of attempts to relate the Financial Times Index of Industrial Ordinary Share Prices to lagged variables of other economic factors, such as Bank Advances, the U.K. Index of Government Securities, the Standard and Poore Index on the New York Stock Exchange, the Bank rate, Sterling liability, and U.K. car production. A measure was also considered to express what was called the Euphoria index, the intention of which was to measure the momentum of the Stock Exchange — if prices have been rising for a long time the market may feel that they will go on rising. (Conversely, we could consider a Dysphoria index, the converse of the Euphoria index, indicating that when the market is falling, investors as a whole may expect the fall to continue.) The appropriate lags on that occasion were determined by examining the cross-correlograms between the F.T. Index and various lagged values of the other variables and taking the points of greatest correlation (or least negative correlation) to indicate the extent of the lag. The series were quarterly and were de-seasonalized by a variation of the Census Mark XI routine. Most of the series contained trend and an attempt was made to remove it, mostly by fitting a linear regression on time, not by moving average.

In economic relationships there is something to be said in favour of using logarithms of the variables, rather than the variables themselves. It may look rather remote from reality to express a price, say, as the sum of a physical quantity of production, an interest rate, and the numbers of miles travelled on railways. Economic equations with this degree of complexity are more realistically expressed in terms of products, as for example in the Cobb—Douglas function. Equivalently, of course, they can be expressed as linear in the logarithms of the quantities involved, provided (as is almost invariably the case) the variables are not negative or small positive quantities. In the study under notice one equation in the original variables and one in the logarithms were computed.

The results were considered by the authors to give sufficiently good forecasts over a period to justify presenting them for general consideration. Some of the equations were surprisingly simple, a regression of the F.T. Index at time t, for example, being related to car production at time $t - 6$ (in quarters), the Financial Times Commodity Index at time $t - 8$, and the Euphoria index at time t, the value of R^2 being 0.95.

11.9   The difficulty with equations of this kind is to validate them. A rejection routine of the kind mentioned in **11.6** and **11.7** may throw out variables which the economist feels must be present; and although, as we mentioned earlier, this may happen in a highly intercorrelated system wherein the variables may substitute one for another, the absence of obviously causal variables and the presence of doubtfully causal ones gives the equation an air of unreality. There is room for a good deal of argument on this point. Box and Newbold (1971) criticized the whole method as being unsound. On the other hand, the history of science teaches us that we should not reject empirically observed relationships just because they cannot be explained in our current state of knowledge. The real test, I suppose, is whether the relationships are stable enough to be used in forecasting — whether, in fact, they "work".

11.10   When we fit an equation over the whole course of a series and judge the excellence of the fit by calculating R^2 we are, in effect, exaggerating its accuracy for forecasting purposes. Had we calculated the equation at various points along the series we should have made a different forecast based on the information we had at those points, and not the information available to us at later points of time. To validate a series retrospectively we should, strictly speaking, go back to each time-point and calculate the equation from such knowledge as we then had (and possibly offering all the variables anew for a rejection routine). This involves more arithmetic than would normally be considered endurable even on a computer. In any case, more conviction is conveyed by prospective than by retrospective validation. But one cannot always wait for a long period to see how a method works. As a compromise, perhaps the most practical procedure is to go back a few units in time and compute the equation for each, so as to see what the forecasting errors would have been.

11.11   The subject is very far from being closed, and it would seem that a good deal more research and experience will be necessary before any firm conclusions can be drawn about this, or any other, forecasting method in particular types of circumstance. We therefore conclude this chapter with a note of some points which the practitioner in Time-Series would wish to take into account in his own work.

(1)   As between the autoprojective and the lagged relationship method, a first consideration might perhaps lead us to expect that the latter would act as a better forecaster because in a sense it uses more information which is relevant to the problem. However, if variables are subject to error (and for some economic variables the errors may be substantial), the importation into the calculations of additional variables may itself introduce extra error which drowns the systematic part of the system. In the engineering phrase, the importation of extra variables may introduce enough noise to obliterate the true message. Obviously, the choice between the two methods will be dependent on individual circumstances and on the amount of observational error in the imported variables.

(2)   The econometric approach referred to in paragraph **11.3** results in what is known as a "reduced form", and the practice has been, until comparatively recently, to estimate each of the equations separately by least squares. In the case of the use of the individual one-equation forecasting method such as we have previously discussed in this chapter, there is no alternative but to fit the one equation which is written down. There are, however, grounds for believing that a bias may be introduced into the estimation, and it remains an open question whether more accurate results would be obtained by the use of several equations (compare Theil, 1971).

(3)   In any case, as we have seen in Example 11.1, the multiple regression coefficient may be a misleading measure of the accuracy of forecasting in the sense that it is highly influenced by trend terms in the series. It is a matter for a somewhat arbitrary choice whether we work on the original series or attempt to remove trend and forecast it separately. There are dangers either way, and we noted in particular in Example 4.4 on page 50 that the variate difference method may give misleading results in the sense that it assumes a model which is in fact not an operative model.

(4)   Although the greater part of the classical theory of regression relies on fitting of least squares, there are doubts whether this is the optimal method in econometric work. The residual terms in equations such as the reduced form of paragraph **11.3** may contain accidentally omitted variables, but even if not, it cannot confidently be assumed that the error terms are normally distributed. It might, for example, be better to fit by minimizing the absolute sum of deviations between expectation and observation, or by some power of the difference other than 2. This admittedly involves further computational problems, but with the aid of an electronic computer such problems are no longer to be regarded as insoluble.

(5)   The use of a rejection routine such as we have described for discarding variables which do not appear to contribute to the goodness of fit may arouse opposition from the point of view of the economist who feels that certain variables ought to appear whatever the rejection routine may suggest. This problem can be overcome on an electronic computer by, so to speak, instructing the machine not to reject certain variables — they are, in the current phrase, "pinned in".

(6)   In an ordinary regression of one variable on p others we are accustomed to think of the number of degrees of freedom in the residual as being $n - p$, n being the sample number. In time-series, as we noted in **1.16**, the observations are in general not independent, and it is hard to see how an exact test of significance of a value of R^2 can be obtained, even if the residuals are normal.

(7)   A further objection which it is possible to launch against the regression method concerns the occasional appearance of long lags. It is sometimes found that an observation of the regressand variable at time t is related to its lagged value or the lagged values of other variables several years previously, and it may be repugnant to common sense to suppose that such long lags are in fact operative. One should not be dogmatic on such points because occasionally it is possible to give an explanation why this should occur. However, for forecasts over relatively short periods of time

there is another possible explanation — namely that, as we noted in earlier chapters, the autoregressive system will oscillate, and the oscillations may not, as it were, have had much time to get out of step, so that the phenomenon presents a spurious cyclicality which may suggest that what looks like a long-lagged relationship is merely a question of history repeating itself.

(8)   Nearly all fitting of such relationships as we have been discussing assumes that the variables are observed without error. If, however, we attempt to write into the model an allowance for errors in variables (as distinct from our former methods which are generally known as errors in equations), some formidable new problems of fitting and estimation appear, even when only a few variables are concerned. For an account of the theory of this subject, reference may be made to Kendall and Stuart's *Advanced Theory of Statistics*, Volume II, Third Edition, Chapter 29.

(9)   Finally, one has to guard against the generation of spurious relationships by aggregation. In Example 1.1 we give an illustration of the effects which may appear. Even if the changes of a series from one point of time to another are more or less random, the aggregation of that series over long periods may generate correlations which are suggestive of effects that are not really present. Unfortunately, as has been mentioned earlier, many of the series which we have to deal with — especially in economics — are aggregated, and there is little that the ordinary practitioner can do to avoid the effect.

11.12   One final remark: it has been the general practice to try out methods of forecasting on series whose true generative mechanism is not known. As we have already remarked, it is not easy to validate the methods against the "true" values except by retrospective or prospective experience. There would seem to be a good deal of room here for testing different methods on artificially constructed series to see how far these methods could correctly indicate the true generating mechanism which, of course, in such cases would be known.

12

Notes on some problems of estimation and significance

12.1 Up to this point we have taken a somewhat intuitive approach to problems of estimation and fitting, assuming for the most part that values calculated from samples are reasonable estimates of parental constants, or that the method of least squares provides a suitable method of fitting curves or regression equations. These assumptions are usually justifiable, though we noted a few cases where dangers were involved in making them — the bias in the estimation of autocorrelations or the sampling variability of spectral ordinates. For the theoretician it is desirable to look into these matters more closely. Unfortunately the problems which arise are quite complicated, even for the simpler cases, and require a mathematical sophistication beyond the scope of this book. Indeed, many of the problems involved in multivariate series have not been solved. In this final chapter, therefore, an attempt is made to give the practitioner some of the results he needs, but he is asked to take the theory on trust.

12.2 Let us first consider why some of the problems are difficult.

(a) We have already noted in Chapter 4 the difficulties of putting error bands round our estimates owing to the fact that "error" may arise from mis-specification of the model. The same situation arises, of course, in ordinary statistical work, where we may have posited the wrong population of origin of a sample; but it is not so severe, partly because we can investigate the robustness of our results under departure of the parent from our assumption, partly because the Central Limit Theorem often confers that robustness on our sampling distributions, and partly because we can often guarantee the randomness of the sample. In time-series, errors of mis-specification are more important, and even where they are absent, exact formulae for standard errors or variances, as we have seen, depend heavily on assumptions of normality or the estimation of a large number of parental parameters. It may be that many of the sampling problems which arise will have to be solved by Monte Carlo methods on the computer rather than by formal mathematics.

(b) Many of the equations which we wish to fit are of the autoregressive or mixed autoregressive cum ordinary regressive type. It does not follow that ordinary least-squares regression theory applies to autoregressive situations; that is to say, the estimators derived by least squares may not be unbiassed or possess minimal variance in the class of linear estimators. Little is known about the exact properties of least-squares estimators for small samples. Fortunately there is a valuable theorem of Mann and Wald (1943, conveniently reproduced in Wald's *Collected Papers*) to the general effect that for large samples classical regression theory applies. More precisely, in the autoregression

$$u_t = -\sum_{j=1}^{k} \alpha_j u_{t-j} + \epsilon_t, \tag{12.1}$$

if the ϵ's are independent (not necessarily normal), then estimators derived by least squares (which are effectively the Yule–Walker equations) are consistent, that is to say the expectation a_i of the parent value α_i converges to α_i as n, the sample size, tends to infinity; and further, the $(a_i - \alpha_i)/n$ are asymptotically distributed in a multivariate normal form with finite dispersion matrix. Moreover, S^2, the mean sum of squares of residuals in the autoregression, is an estimator of var ϵ, and the covariance of two estimators a_j, a_l is estimated by S^2 multiplied by the inverse of the matrix $(\Sigma\, u_{t-j}\, u_{t-l})/n$.

(c) In default of more exact knowledge we therefore assume that, in general, least-squares theory applies to autoregression. Indeed, we go further and assume that it applies to mixed regression systems as, in effect, we did in Chapter 11. However, when we have a set of such equations constituting a model, the treatment of each equation separately by least-squares methods without modification can lead to some important biasses. The treatment of such systems is still imperfectly understood. Reference may be made to Fisk (1967) and Theil (1971) for accounts of the subject in the rather wider context of model-building.

(d) A number of the models which we use in time-series analysis, even of the single-equation type, suppose that the residual ϵ is independent from one sample to another and in particular from one term to the next in a temporally arranged set. The adequacy or accuracy of this hypothesis is not easy to test, especially for autoregression.

12.3   The kind of schemes whose parameters we have to estimate or the adequacy of whose fit we have to test may be arranged in a kind of ascending hierarchy of complexity.

(a) First of all, the autoregressive scheme of type (12.1) with random residuals.

(b) Autoregressions with residuals which are not random, and in particular those with residuals which are moving averages of random terms:

$$\sum_{j=0}^{k} \alpha_j u_{t-j} = \sum_{j=0}^{l} \beta_j \epsilon_{t-j}. \tag{12.2}$$

(c) Regressions with residuals which are autoregressive:

$$y_t = \sum_{j=1}^{q} \beta_j x_{jt} + u_t. \tag{12.3}$$

(d) Mixed systems in which some of the x's in (12.3) are lagged values of the y's, but with random residuals:

$$y_t = -\sum_{j=1}^{k} \alpha_j y_{t-j} + \sum_{l=1}^{q} \beta_l x_{lt} + \epsilon_t. \tag{12.4}$$

(e) The similar situation where the residuals are moving averages of random terms or are autoregressive.

(f) Systems involving several equations of the foregoing types.

Fitting autoregressions

12.4   For schemes of type (12.1) the Yule–Walker equations should be used. The only problem is to decide what order of scheme to fit. It is natural to go on extending the number of terms k until the fit does not materially improve.

Table 12.1   *Residual values of the sheep series of Table 1.2 after removal of trend by a simple 9-point moving average*

Year	Residual (10 000)	Year	Residual (10 000)	Year	Residual (10 000)
1871	− 176	1893	+ 34	1915	+ 19
72	− 112	94	− 103	16	+ 128
73	+ 50	95	− 104	17	+ 97
74	+ 141	96	− 15	18	+ 69
75	+ 60	97	− 23	19	− 29
76	− 20	98	+ 17	20	− 174
77	+ 12	99	+ 71	21	− 107
78	+ 82	1900	+ 35	22	− 142
79	+ 130	01	+ 16	23	− 109
80	− 14	02	− 27	24	− 23
81	− 166	03	− 32	25	+ 60
82	− 179	04	− 49	26	+ 121
83	− 84	05	− 61	27	+ 94
84	+ 38	06	− 52	28	− 25
85	+ 97	07	− 24	29	− 90
86	+ 8	08	+ 68	30	− 75
87	− 5	09	+ 141	31	+ 72
88	− 105	10	+ 119	32	+ 152
89	− 99	11	+ 66	33	+ 112
90	+ 35	12	− 52	34	− 64
91	+ 159	13	− 117	35	− 87
92	+ 167	14	− 61		

Example 12.1

Table 12.1 shows the residuals obtained from the sheep series of Table 1.2 on removal of trend by a simple 9-point moving average. The first 10 serial correlations of this series are as follows:

Order of correlation		Order of correlation	
1	0·595	6	0·144
2	−0·151	7	0·203
3	−0·601	8	0·118
4	−0·537	9	0·006
5	−0·138	10	−0·078

$$(12.5)$$

We have noticed in **6.23** that for a scheme of order k the partial correlation of terms more than k apart should vanish. The following are the partials:

lag k	Partial r of lag k	$\Pi(1-r^2) = 1-R^2$
1	0·595	0·6460
2	−0·782	0·2509
3	0·097	0·2485
4	− 0·183	0·2402
5	0·031	0·2400
6	0·014	0·2400

$$(12.6)$$

Here, for example, the value of partial r for $k = 3$ is the partial correlation of u_t and u_{t+3} when their dependence on the intervening terms u_{t+1}, u_{t+2} is removed. The column on the right gives the residual variance (divided by var u), which is equal to the complement of the multiple correlation coefficient R^2. Thus, for $k = 3$, a three-term equation represents all but 24·85 per cent. of the variation. It is evident that a scheme of order 2 (a Yule scheme), with a value of $1 - R^2$ equal to 0·2509, is as far as we need to go. Further terms add little or nothing to the representation.

The estimates of the constants α_1 and α_2 in the series are obtained from (6.24) and (6.25) as

$$-a_1 = \frac{r_1(1-r_2)}{1-r_1^2} = 1\cdot060, \qquad -a_2 = \frac{r_2-1}{1-r_1^2}+1 = -0\cdot782,$$

and the autoregressive equation is

$$u_t = 1\cdot060u_{t-1} - 0\cdot782u_{t-2} + \epsilon_t. \qquad (12.7)$$

12.5   Quenouille (1947) used the property of the partial correlations to provide a test of fitting autoregressive schemes which avoids the explicit calculation of the partials. Consider a variable η_t defined by

$$\sum_{j=0}^{k} \alpha_j u_{t+j} = \eta_t, \tag{12.8}$$

where, in distinction from (12.7), the suffixes of the u's increase with j. Then

$$\text{cov}\,(\eta_t, \eta_{t+l}) = \text{E}\,\{\Sigma\,\alpha_i u_{t+i}\,\Sigma\,\alpha_j u_{t+j+l}\}$$

$$= \sum_{i,j=0}^{k} \alpha_i \alpha_j\,\gamma_{|l+j-i|}, \tag{12.9}$$

where γ_p is the pth autocovariance of u_t,

$$= \sum_{i=0}^{k} \alpha_j \sum_{j=0}^{k} \alpha_j\,\gamma_{|j+k-l|}. \tag{12.10}$$

The second part of the summation vanishes in virtue of the Yule–Walker equations, for $l > 0$, and the same result holds for $l < 0$. Hence

$$\left.\begin{array}{c}\text{cov}\,(\eta_t, \eta_{t+l}) = 0 \\[2mm] \text{var}\,\eta_t = \text{var}\,\epsilon_t.\end{array}\right\} \tag{12.11}$$

It is to be noticed that η_t depends on ϵ_{t+k} and previous ϵ's, and is therefore independent of ϵ_{t+k+l} for $l > 0$. Now from (12.8) η_t is the residual of u_t after the removal $u_{t+1}, \ldots, u_{t+k}$; whereas ϵ_{t+j} is u_{t+j} after the removal of $u_{t+j-1}, \ldots, u_{t+j-k}$. Hence, for $j > k$ the correlation between η_t and ϵ_{t+j} is the partial correlation of terms in the series distance j apart. Define then

$$q_j = \frac{1}{n} \sum_{t=1}^{n} \epsilon_{t+j}\,\eta_t. \tag{12.12}$$

We have

$$\text{E}\,(q_j) = 0, \quad j > k \tag{12.13}$$

$$\text{var}\,q_j = \text{E}\,(\epsilon_{t+j}^2)\,\text{E}\,(\eta_t^2)$$

$$= (\text{var}\,\epsilon)^2, \quad j > k. \tag{12.14}$$

$$\text{cov}\,(q_i, q_{i+l}) = 0, \quad l \neq 0. \tag{12.15}$$

Standardizing the quantities q, we define

$$w_j = \frac{q_j}{\text{var}\,u}$$

$$= \frac{1}{n\,\text{var}\,u} \sum_{t=1}^{n} \epsilon_{t+j}\,(\alpha_0 u_t + \ldots + \alpha_k u_{t+k}), \quad j > k, \tag{12.16}$$

and each w has zero mean, variance equal to $(\text{var}\,\epsilon/\text{var}\,u)^2$, and is uncorrelated with the other w's.

For large samples, we have from (12.16)

$$w_j = \frac{1}{n \, \text{var} \, u} \, \text{E} \{(\alpha_0 u_{t+j} + \ldots + \alpha_k u_{t+j-k})(\alpha_0 u_t + \ldots + \alpha_k u_{t+k})\}$$

$$= A_0 r_j + A_1 r_{j-1} + A_2 r_{j-2} + \ldots + A_{2k} r_{j-2k}, \tag{12.17}$$

where the A's are given by

$$A_i = \sum_{j=0}^{i} \alpha_j \alpha_{i-j}. \tag{12.18}$$

Example 12.2

For the Yule scheme we find

$$A_0 = 1, \quad A_1 = 2\alpha_1, \quad A_2 = \alpha_1^2 + 2\alpha_2, \quad A_3 = 2\alpha_1\alpha_2, \quad A_4 = \alpha_2^2$$

and hence, asymptotically,

$$w_j = r_j + 2\alpha_1 r_{j-1} + (\alpha_1^2 + 2\alpha_2)r_{j-2} + 2\alpha_1\alpha_2 r_{j-3} + \alpha_2^2 r_{j-4}, \quad j > 2, \tag{12.19}$$

and this has variance (cf. equation (6.49))

$$\frac{1}{n}\left[\frac{1-\alpha_2}{1+\alpha_2}\{(1+\alpha_2)^2 - \alpha_1^2\}\right]. \tag{12.20}$$

For instance, in the sheep series of Example 12.1 we find, substituting $\alpha_1 = -1\cdot060$, $\alpha_2 = +0\cdot782$ in (12.19),

$$w_j = r_j - 2\cdot120 r_{j-1} + 2\cdot683 r_{j-2} - 1\cdot658 r_{j-3} + 0\cdot612 r_{j-4}.$$

We find also $w_3 = 0\cdot025$, $w_4 = 0\cdot043$, $w_5 = 0\cdot001$, with a standard error of $0\cdot031$, which confirms our finding that a second-order (Yule) scheme will account for the process.

Rosenhead (1968) has generalized Quenouille's test to multivariate series.

12.6   With a computer and a routine for adding and subtracting variables it is possible to fit autoregressions more directly, provided that the variables can be brought in in the right order. If we submit data to a stepwise method or an optimal regression method the variables may be imported in a different order. This is what happened in Example 11.2, where the routine brought in I_{t-1} but rejected further terms up to I_{t-6} and then further terms up to I_{t-13}. There seems to be no objection to this, and there is certainly none so far as concerns the least-squares criterion of fit which forms the basis of the use of R^2. The physical interpretation of such systems offers more of a problem, for the systematic part of the series then depends on values at isolated points in the past and not on the intervening values. There may well be explanations of the phenomenon in particular cases.

Moving-average schemes

12.7   Since the moving-average on the right in (12.2) is an average of finite extent, whereas the solution of an autoregressive scheme involves an average of infinite extent, it might be thought that moving-average schemes would be easier to handle than autoregressions. The reverse is the case. Consider even such a simple scheme as

$$u_t = \epsilon_t + \beta\epsilon_{t-1}. \tag{12.21}$$

Since

$$\text{var } u_t = (1 + \beta^2) \text{ var } \epsilon$$

$$\text{cov } (u_t, u_{t+1}) = \beta,$$

it seems plausible to estimate β as a root of

$$\rho_1 = \frac{\beta}{1 + \beta^2}, \tag{12.22}$$

where ρ_1 is the first autocorrelation. This is certainly possible, but, as Whittle (1953) showed, the resulting estimator is very inefficient, that is to say, there are estimators with a much smaller variance. For example, with $\beta = \frac{1}{2}$, the variance of an estimator based on (12.22) is more than three times the optimal value.

12.8   We noticed this topic in **9.22** in discussing the Box–Jenkins model for short-term forecasting. Perhaps in the future all complicated estimational problems of this kind will be solved by iterative routines on the computer. It is, however, worth while noting, as Durbin (1959) pointed out, that the problem of the pure moving-average scheme

$$u_t = \sum_{j=0}^{q} \beta_j \epsilon_{t-j} \tag{12.23}$$

can be solved by turning it into an infinite autoregression. Consider, in fact, (12.21) expressed in the equivalent form

$$u_t - \beta u_{t-1} + \beta^2 u_{t-2} - \beta^3 u_{t-3} + \ldots = \epsilon_t. \tag{12.24}$$

Let us stop the series on the left at some point $t - k$. We then have a particular kind of autoregression with $\alpha_j = (-\beta)^j$.

The error committed is the remaining part of the series

$$(-\beta)^{k+1} u_{t-k-1} + (-\beta)^{k+2} u_{t-k-2} + \ldots, \text{etc.}$$

$$= (-\beta)^{k+1} \epsilon_{t+k-1}.$$

This is small for $|\beta| < 1$, having a variance of β^{2k+2} var ϵ.

Treating the problem as a finite autoregression gives us estimates of the powers $(-\beta)$, $(-\beta)^2$, etc. as an array of values $a_1, a_2, \ldots, a_k$. These have to be reconciled. By an appeal to the Mann–Wald theorem, Durbin showed that a maximum likelihood estimator b of β is given by

$$b = -\sum_{j=0}^{k-1} a_j a_{j+1} \bigg/ \sum_{j=0}^{k} a_j^2 \tag{12.25}$$

and that this has the optimal (minimum) variance $(1 - \beta^2)/n$.

The extension of these results to the more general moving-average scheme (12.23) is possible but involves some formidable algebra.

For mixed autoregressive moving-average schemes of type (12.2) the best procedure to obtain arithmetic solutions appears to be some iterative routine such as that described briefly in 9.22.

Regression with autocorrelated residuals

12.9   Consider now a regression

$$y_t = \sum_{j=1}^{q} \beta_j x_{jt} + u_t, \tag{12.26}$$

where u_t is itself autocorrelated according to

$$\sum_{j=0}^{k} \alpha_j u_{t-j} = \epsilon_t. \tag{12.27}$$

If the α's were known, we could substitute for u from (12.26) in (12.27) to obtain

$$y_t' = \sum_{l=1}^{q} \beta_l x_{lt}' + \epsilon_t, \tag{12.28}$$

where

$$y_t' = \sum_{j=0}^{k} \alpha_j y_{t-j} \tag{12.29}$$

$$x_{jt}' = \sum_{j=0}^{k} \alpha_j x_{l, t-j}. \tag{12.30}$$

(12.28) is now an ordinary regression. The difficulty is that we do not know the α's or the extent of the autoregression. As on previous occasions of a similar kind, it appears necessary to proceed by some iterative process. Cochrane and Orcutt (1949) suggested guessing the values of α, estimating β's from (12.28), reconsidering the u's obtained from (12.26) with these estimates of β, and so on, hopefully to convergence on a scheme with random ϵ's. Durbin (1960) suggested an alternative based on maximum likelihood, but the expressions are not easy to handle.

12.10   If some of the x's are lagged values of the y's, so that we have a mixed autoregressive–regressive situation,

$$y_t = -\sum_{j=1}^{k} \alpha_j y_{t-j} + \sum_{l=1}^{q} \beta_l x_{lt} + \epsilon_t, \tag{12.31}$$

the foregoing theory does not apply, much as we noticed that ordinary least-squares theory does not apply to autoregressions. As an extension of the Mann–Wald theorem, Durbin has shown (1960) that asymptotically the properties of least-squares estimators in such a system are the same as those without lagged variables, whether or not the residuals are normally distributed. In the present state of knowledge, then, it seems best, as a practical procedure, to treat such equations by least squares as if they were ordinary regressions.

Analysis of residuals

12.11 For any kind of fitting of time-series models it is obviously important to scrutinize the residuals, i.e. the differences between the observed and the fitted values. We noted in considering moving averages in Chapter 3 that the averaging process distorted the residuals to some extent. The same effect is present when we fit a polynomial over the course of the entire series, and indeed in any fitting of the regression type.

Nevertheless it is very useful (and with a computer is no longer an arithmetical nuisance) to work out all the observed residuals and not, as in precomputer days, to rely largely on the calculation of their sums of squares. Notwithstanding that the observed residuals differ to some extent from the real residuals, important effects which may impair the fitting may show up in the observed set, especially in time-series. For example, if there are any exceptional outlying observations, their existence will, as a rule, be revealed by a large residual term. Or again, a run of positive or negative signs in the series of residuals is some indication that the model which has been fitted to them is inadequate.

12.12 Suppose that we have a regression relationship between a regressand variable y and regressors $x_1, x_2, \ldots, x_p$,

$$y = \sum_{j=0}^{p} \beta_j x_j + \epsilon, \tag{12.32}$$

or in matrix form with n observations and y as an $n \times 1$ matrix,

$$\mathbf{y} = \mathbf{X}\boldsymbol{\beta} + \boldsymbol{\epsilon}. \tag{12.33}$$

The quantities of interest here are the ϵ's. However, we do not observe them. Estimating the β's by least squares gives us

$$\mathbf{b} = (\mathbf{X}'\mathbf{X})^{-1}\mathbf{X}'\mathbf{y}, \tag{12.34}$$

and consequently the observed residuals, ϵ say, are given by

$$\mathbf{e} = \mathbf{y} - \mathbf{X}\mathbf{B} = \mathbf{y} - \mathbf{X}(\mathbf{X}'\mathbf{X})^{-1}\mathbf{X}'\mathbf{y}$$
$$= \{\mathbf{I}_n - \mathbf{X}(\mathbf{X}'\mathbf{X})^{-1}\mathbf{X}'\}\mathbf{y}, \tag{12.35}$$

where $\mathbf{I}_n$ is the unit matrix of order n,

$$= \{\mathbf{I}_n - \mathbf{X}(\mathbf{X}'\mathbf{X})^{-1}\mathbf{X}'\}\{\mathbf{X}\mathbf{B} + \boldsymbol{\epsilon}\}$$
$$= \{\mathbf{I}_n - \mathbf{X}(\mathbf{X}'\mathbf{X})^{-1}\mathbf{X}'\}\boldsymbol{\epsilon}. \tag{12.36}$$

Thus the observed residuals $\mathbf{e}$ are linear functions of the real residuals $\boldsymbol{\epsilon}$. If we write

$$\mathbf{M} = \mathbf{I} - \mathbf{X}(\mathbf{X}'\mathbf{X})^{-1}\mathbf{X}', \qquad (12.37)$$

then $\mathbf{e} = \mathbf{M}\boldsymbol{\varepsilon}$. Unfortunately the relation is not invertible, for $\mathbf{M}$ is degenerate, as may be seen by postmultiplying (12.37) by $\mathbf{X}$.

The Durbin–Watson test

12.13   More important, perhaps, is the fact that serial correlations are generated in the observed residuals, which makes it impossible to test the original model for independence of residuals merely by testing the observed residuals for independence. A more sophisticated test is required and this has been provided by Durbin and Watson (1971 and previous papers).

The Durbin–Watson statistic d is defined in terms of the observed residuals by

$$d = \frac{\sum_{i=2}^{n} (e_i - e_{i-1})^2}{\sum_{i=1}^{n} e_i^2}. \qquad (12.38)$$

This, apart from the end effect, is $2(1 - r_1)$, where r_1 is the first serial correlation of the observed residuals. Alternatively it can be looked on as the sum of squares of first differences of the residuals divided by their sum of squares, as in the variate-difference method. If the residuals are highly positively correlated the value of d is near zero; if they are uncorrelated it is near 2.

12.14   The actual distribution of d, which is necessary for an exact test of the hypothesis that the real residuals are uncorrelated, is a very complicated one. It was shown, however, that d lies between two statistics d_L and d_U whose distributions are related to those of R.L. Anderson's distribution of the first serial correlation coefficient (cf. **7.6**). The tables in Appendix B give the significance points of d_L and d_U for the 1 per cent. and the 5 per cent. levels, for $n = 15$ to 100, and k' (the number of regressors in the regression equation apart from the constant term) from 1 to 5. If an observed d falls below the tabulated value of d_L we reject the hypothesis that the original residuals are uncorrelated, and if it falls above the tabulated value of d_U the hypothesis is accepted. For cases of indecision where d falls between d_L and d_U various procedures have been suggested. The most recent one, favoured by Durbin and Watson, is to approximate to the distribution of d by a statistic based on d_U, namely

$$d^* = a + bd_U, \qquad (12.39)$$

where a and b are constants chosen so that d^* and d have the same mean and variance. The formulae are fairly complicated. Reference may be made to Durbin and Watson (1971) for the details. Doubtless machine routines will be written for practical use. They already exist for the calculation of d.

12.15 The exactitude of the test depends on an assumption of normality in the distribution of the residuals ϵ. There is some evidence that Anderson's distribution is fairly robust under moderate departures from normality, so that presumably the Durbin–Watson test is also reasonably robust. It applies to those cases where we fit a regression over the whole course of the series. Strictly speaking it does not apply to moving averages, although it would apply to each part of the moving average, when obtained by least-squares methods.

12.16 A more serious drawback is that the Durbin–Watson test does not apply without modification to autoregressive series. A discussion in the large-sample case was given by Durbin himself (1970), particularly for the type of scheme with a residual which is itself of Markoff autoregressive type. For example, with a scheme of type (12.4) in which the residuals are related by

$$\epsilon_t = \rho \epsilon_{t-1} + \eta_t, \tag{12.40}$$

η_t being an independent normal sequence, the statistic g, which is $1 - \frac{1}{2}d$ and is therefore an estimator of the first serial correlation of residuals, can be used to provide a test of the parent ρ. We construct

$$h = g \left(\frac{n}{1 - n \, \text{var} \, a_1} \right)^{\frac{1}{2}}, \tag{12.41}$$

where var a_1 is the estimated variance of α_1 given by ordinary least-squares theory in equation (12.4) with $k = 1$. This may, for large samples, be tested as an ordinary standard normal variate. More detailed results have not, apparently, been worked out and in any case the position for small samples remains relatively unexplored. The general message, however, is plain. The Durbin–Watson test, which applies to ordinary regressions (where there would be no term in var a_1) is vulnerable to sampling variation in the estimators of the autoregressive parameters.

NOTES

(1) For the analysis of residuals when regression relationships are fitted to time-series reference should be made to an important paper by Brown, R.L., Durbin, J. and Evans, J.M., (1975), Techniques for testing the constancy of regression relationships over time. *J.R. Statist. Soc.*, B, **37**.

(2) For a simple test of residuals see Geary, R.C. (1970), Relative efficiency of count of sign changes for assessing residual autoregression in least-squares regression. *Biometrika*, **57**, 123.

REFERENCES

Anderson, R.L. (1942), Distribution of the serial correlation coefficient, *Ann. Math. Statist.*, **13**, 1.

Anderson, T.W. (1958), *The Statistical Analysis of Time-Series*. John Wiley & Sons, London.

Bartlett, M.S. (1946), On the theoretical specification and sampling properties of auto-correlated time-series. *Suppl. J. R. Statist. Soc.*, **8**, 27, 85 and (1948) **10**, 85.

Bartlett, M.S. (1950), Periodogram analysis and continuous spectra. *Biometrika*, **37**, 1.

Bartlett, M.S. (1955), *An Introduction to Stochastic Processes, with special reference to Methods and Applications*. Cambridge University Press.

Beale, E.M.H., Kendall, M.G. and Mann, D.W. (1967), The discarding of variables in multivariate analysis. *Biometrika*, **54**, 357.

Beveridge, W.H. (1921), Weather and harvest cycles. *Econ. J.*, **31**, 429.

Birkhoff, G.D. (1931), Proof of the ergodic theorem. *Proc. Nat. Acad. Sci.*, **17**, 656.

Blackman, R.B. and Tukey, J. (1958), *The Measurement of Power Spectra from the point of view of Communication Engineering*. Dover Publications, New York.

Bliss, C.I. (1958), Periodic Regression in Biology and Climatology. Bulletin No. 615, Connecticut Agricultural Experiment Station, New Haven.

Box, G.E.P. and Jenkins, G.M. (1970), *Time-Series Analysis, Forecasting and Control*. Holden-Day (McGraw–Hill Book Co., New York and Maidenhead, Eng.).

Box, G.E.P. and Newbold, P. (1971), Some comments on a paper of Coen, Gomme and Kendall. *J. R. Statist. Soc., A*, **134**, 299.

Brown, R.G. (1963), *Smoothing, Forecasting and Prediction*. Prentice-Hall, Englewood Cliffs, N.J.

Burman, J.P. (1965), Moving seasonal adjustment of economic time series. *J. R. Statist. Soc., A*, **128**, 534 and (1966) **129**, 274.

Cochrane, D. and Orcutt, G.H. (1949), Application of least-squares regression to rela-tionships containing autocorrelated error terms. *J. Amer. Statist. Ass.*, **44**, 32.

Coen, P.G., Gomme, E.D. and Kendall, M.G. (1969), Lagged relationships in economic forecasting. *J. R. Statist. Soc., A*, **132**, 133.

Cowden, D.J. (1962), *Weights for fitting polynomial secular trends*. Technical Paper No. 4, School of Business Administration, University of North Carolina.

Daniell, P.J. (1946), Discussion on "Symposium on Autocorrelation in Time-series". *Supp. J. R. Statist. Soc.*, **8**, 88.

Daniels, H.E. (1956), The approximate distribution of serial correlation coefficients. *Biometrika*, **43**, 169.

Daniels, H.E. (1962), The estimation of spectral densities. *J. R. Statist. Soc., B*, **24**, 185.

Daniels, H.E. (1970), Autocorrelation between first-differences of mid-ranges. *Econometrika*, **34**, 215.

Dewey, E.R. (1963), The 18·2-year cycle in immigration, U.S.A., 1820–1962. Foundation for the Study of Cycles, Inc., Pittsburgh, Pa.

Dixon, W.J. (1944), Further contributions to the problem of serial correlation. *Ann. Math. Statist.*, **15**, 119.

Durbin, J. (1959), Efficient estimation of parameters in moving-average models. *Biometrika*, **46**, 306.

Durbin, J. (1960), Estimation of parameters in time-series regression models. *J. R. Statist. Soc., B*, **22**, 139.

Durbin, J. (1961), Efficient fitting of linear models for continuous stationary time series from discrete data. *Bull. Int. Statist. Inst.*, **38**, (4), 273.

Durbin, J. (1963), Trend elimination for the purpose of estimating seasonal and periodic components of time-series. In the book edited by Rosenblatt, *q.v.*

Durbin, J. (1970), Testing for serial correlation in least-squares regression when some of the regressors are lagged dependent variables. *Econometrica*, **38**, 410.

Durbin, J. and Watson, G.S. (1950, 1951, 1971), Testing for serial correlation in least-squares regression. *Biometrika*, **37**, 409; **38**, 159; **58**, 1.

Fisk, P.R. (1967), *Stochastically Dependent Equations*. Charles Griffin & Co., London.

Foster, F.G. and Stuart, A. (1954), Distribution-free tests in time-series based on the breaking of records. *J. R. Statist. Soc., B,* **16**, 1.

Funkhauser, H.G. (1936), A note on a 10th century graph. *Osiris*, 1, Bruges.

Gleissberg, W. (1945), Eine Aufgabe der Kombinatorik in der Wahrscheinlichkeitsrechnung. *Univ. Istanbul Rev. Fac. Sci., A,* **10**, 25.

Granger, C.W.J. (1963), The effect of varying month-length on the analysis of economic time-series. *L'Industria,* **1**, 3, Milano.

Granger, C.W.J. and Hatanaka, M. (1964). *Spectral Analysis of Economic Time-Series.* Princeton Univ. Press, Princeton, N.J.

Granger, C.W.J. and Hughes, A.O. (1971), A new look at some old data: the Beveridge wheat price series. *J. R. Statist. Soc., A,* **134**, 413.

Grenander, U. and Rosenblatt, H.M. (1957), *Statistical Analysis of Stationary Time Series.* John Wiley & Sons, New York, & Chichester, Eng.

Gudmundsson, G. (1971), Time-series analysis of imports, exports and other economic variables. *J. R. Statist. Soc., A,* **134**, 383.

Hannan, E.J. (1960a), *Time Series Analysis.* Methuen, London.

Hannan, E.J. (1960b), The estimation of seasonal variation. *Australian J. Statist.,* **2**, 1.

Harrison, P.J. (1965), Short-term sales forecasting. *Applied Statistics,* **14**, 102.

Harrison, P.J. (1967), Exponential smoothing and short-term sales forecasting. *Mgmt. Sci.,* **13**, 821.

Harrison, P.J. and Stevens, C.F. (1971), A Bayesian approach to short-term forecasting. *Opl. Res. Q.,* **22**, 341.

Holt, C.C. (1957), *Forecasting seasonals and trends by exponentially weighted moving averages.* Carnegie Inst. Tech. Res. Mem. No. 52.

Jevons, W.S. (1879), *The Principles of Science.* London.

Kalman, R.E. and Bucy, R.S. (1961), New results in linear filtering and prediction theory. *Trans. A.S.M.E.: Journal of Basic Engineering,* p. 95.

Kendall, M.G. (1945), On the analysis of oscillatory time-series. *J. R. Statist. Soc., A,* **108**, 93.

Kendall, M.G. (1946), *Contributions to the Study of Oscillatory Time-series.* Cambridge University Press.

Kendall, M.G. (1949), The estimation of parameters in lower autoregressive time-series. *Econometrica, Supp.,* **17**, 44.

Kendall, M.G. (1969), *Rank Correlation Methods,* 4th edn. Charles Griffin & Co., London.

Kendall, M.G. (1971), Studies in the history of probability and statistics, XXVI: The work of Ernst Abbe. *Biometrika,* **58**, 369.

Kendall, M.G. and Stuart, A. (1968), *Advanced Theory of Statistics*, Vol. 3, 2nd edn. Charles Griffin & Co., London.

Khintchine, A. Ya. (1932), Zu Birkhoffs Lösung des Ergodenproblems. *Math. Ann.*, **107**, 485.

King, P.D. (1956), Increased frequency of births in the morning hours. *Science*, **123**, 985.

Leipnik, R.P. (1947), Distribution of the serial correlation coefficient in a circularly correlated universe. *Ann. Math. Statist.*, **18**, 80.

Levene, H. (1952), On the power function of tests of randomness based on runs up and down. *Ann. Math. Statist.*, **23**, 34.

Madow, W.G. (1945), Note on the distribution of the serial correlation coefficient. *Ann. Math. Statist.*, **16**, 308.

Mann, H.B. (1945), Nonparametric tests against trend. *Econometrica*, **13**, 245.

Mann, H.B. and Wald, A. (1943), On the statistical treatment of linear stochastic difference equations. *Econometrica*, **11**, 173.

Moore, G.H. and Wallis, W.A. (1943), Time series significance tests based on signs of differences. *J. Amer. Statist. Ass.*, **38**, 153.

Moran, P.A.P. (1947), Some theorems on time-series, I. *Biometrika*, **34**, 281.

Moran, P.A.P. (1948), Some theorems on time-series, II. *Biometrika*, **35**, 255.

Moran, P.A.P. (1967), Testing for serial correlation with exponentially distributed variates. *Biometrika*, **54**, 395.

Nold, F.C. (1972), *A bibliography of applications of techniques of spectral analysis to economic time-series.* Technical Report No. 66, Institute for Mathematical Studies in the Social Sciences, Stanford University, California.

Parzen, E. (1961), Mathematical considerations in the estimation of spectra: Comments on the discussions of Messrs Tukey and Goodman. *Technometrics*, **3**, 167, 232.

Playfair, W. (1821), *A Letter on our Agricultural Distress.* London.

Quenouille, M.H. (1947), A large-sample test for the goodness of fit of autoregressive schemes. *J. R. Statist. Soc.*, **110**, 123.

Quenouille, M.H. (1949), A method of trend elimination. *Biometrika*, **36**, 75.

Quenouille, M.H. (1953), Modifications to the variate-difference methods. *Biometrika*, **40**, 383.

Quenouille, M.H. (1956), Notes on bias in estimation. *Biometrika*, **43**, 353.

Quenouille, M.H. (1957, 1968), *The Analysis of Multiple Time-Series.* Charles Griffin & Co., London.

Quenouille, M.H. (1958), Discrete autoregressive schemes with varying time-intervals. *Metrika*, **1**, 21.

Reid, D.J. (1971), Forecasting in action: A comparison of forecasting techniques in economic time-series. *Joint Conference of O.R. Society's Group on Long Range Planning and Forecasting.*

Rhodes, E.C. (1921), *Smoothing. Tracts for Computers, No. 6.* Cambridge University Press.

Rosenberg, B. (1970), The distribution of mid-range – a comment. *Econometrica*, **38**, 176.

Rosenblatt, M. (ed.) (1963), *Proceedings of the Symposium of Time-Series Analysis held at Brown University, 1962.* John Wiley & Sons, New York, & Chichester, Eng.

Rosenhead, J. (1968), An extension of Quenouille's test for the compatibility of correlation structures in time. *J. R. Statist. Soc., B*, **30**, 180.

Sargan, J.D. (1953). An approximate treatment of the properties of the correlogram and periodogram. *J. R. Statist. Soc., B*, **15**, 140.

Shiskin, J. (1967), The X–11 Variant of the Census Method II Seasonal Adjustment Program. Technical Paper No. 15, U.S. Bureau of the Census.

Slutzky, E. (1927), The summation of random causes as the source of cyclic processes. (Russian). English trans., 1937. *Econometrica, 5*, 105.

Spencer, J. (1904). On the graduation of the rates of sickness and mortality. *J. Inst. Act., 38*, 334.

Stuart, A. (1954), Asymptotic relative efficiencies of distribution-free tests of randomness against normal alternatives. *J. Amer. Statist. Ass., 49*, 147.

Stuart, A. (1956), The efficiencies of tests of randomness against normal regression. *J. Amer. Statist. Ass., 51*, 285.

T.A.S.C. (The Analytic Sciences Corporation) (1971), *A short course on Kalman Filter Theory and Application.* T.A.S.C., Reading, Mass.

Theil, H. (1971), *The Principles of Econometrics.* John Wiley & Sons, New York, & Chichester, Eng.

U.S. Bureau of the Census (1965), A Spectral Study of "Overadjustment" for Seasonality. Working Paper No. 21. Washington, D.C.

Wallis, W.A. and Moore, G.H. (1941), A significance test for time-series analysis. *J. Amer. Statist. Ass., 36*, 401.

Whittle, P. (1953), The analysis of multiple time-series. *J. R. Statist. Soc., B, 15*, 125.

Wilson, L.L. (1964), *Catalogue of Cycles. Part I – Economics.* Foundation for the Study of Cycles, Inc., Pittsburgh, Pa.

Winters, P.R. (1960), Forecasting sales by exponentially weighted moving averages. *Mgmt. Sci., 6*, 324.

Wold, H.O. (1954), *A Study in the Analysis of Stationary Time-Series,* 2nd edn. Almqvist & Wicksell, Uppsala.

Wold, H.O. (1965) (ed.), *Bibliography on Time-Series and Stochastic Processes.* Oliver and Boyd, Edinburgh.

Working, H. (1960), Note on the correlation of first differences of averages in a random chain. *Econometrica, 28*, 916.

Yule, G. Udny (1927), On a method of investigating periodicities in disturbed series, with special reference to Wolfer's sunspot numbers. *Phil. Trans., A, 226*, 267.

Yule, G. Udny (1971), *Statistical Papers: Selected by Alan Stuart and M.G. Kendall.* Charles Griffin & Co., London.

Weights for fitting polynomial trends

The following tables are extracted by permission from Dudley J. Cowden, *Weights for fitting polynomial secular trends*, Technical Paper No. 4, School of Business Administration, University of North Carolina, 1962. Professor Cowden gives values up to N (the extent of the moving average) = 25, and also values for even N.

Except in one or two early tables, the tables give the weights required to fit at one end of the series, those for the other end being given by symmetry. For example, for fitting a straight line (a simple moving average) to nine points the weights for the first point are $\frac{1}{45}$ [17, 14, 11, 8, 5, 2, -1, -4, -7]. Those for the second are $\frac{1}{180}$ [56, 47, 38, 29, 20, 11, 2, -7, -16] and so on. Conversely, for the last four points, the weights are (reading the table upwards) for the last but three, [8, 11, 14, 17, 20, 23, 26, 29, 32] and so on.

The columns headed 0 give the weights required to extrapolate the fitting one unit beyond the end of the observed series.

The sums in the last row but one are the sums of the integral weights given in the table.

The final row in the table is the square root of the error-reducing factor, i.e. the square root of the sum of the squares of the weights.

LINEAR

N = 3

0	1	2	1	0
4	5	1	−1	−2
1	2	1	2	1
−2	−1	1	5	4
3	6	3	6	3
1·528	·913	·577	·913	1·528

N = 5

0	1	2	3	2	1	0
8	3	4	1	0	−1	−4
5	2	3	1	1	0	−1
2	1	2	1	2	1	2
−1	0	1	1	3	2	5
−4	−1	0	1	4	3	8
10	5	10	5	10	5	10
1·049	·775	·548	·447	·548	·775	1·049

N = 7

0	1	2	3	4
4	13	5	7	1
3	10	4	6	1
2	7	3	5	1
1	4	2	4	1
0	1	1	3	1
−1	−2	0	2	1
−2	−5	−1	1	1
7	28	14	28	7
·845	·681	·535	·423	·378

N = 9

0	1	2	3	4	5
16	17	56	22	32	1
13	14	47	19	29	1
10	11	38	16	26	1
7	8	29	13	23	1
4	5	20	10	20	1
1	2	11	7	17	1
−2	−1	2	4	14	1
−5	−4	−7	1	11	1
−8	−7	−16	−2	8	1
36	45	180	90	180	9
·726	·615	·511	·422	·357	·333

N = 11

0	1	2	3	4	5	6
20	7	15	25	10	15	1
17	6	13	22	9	14	1
14	5	11	19	8	13	1
11	4	9	16	7	12	1
8	3	7	13	6	11	1
5	2	5	10	5	10	1
2	1	3	7	4	9	1
−1	0	1	4	3	8	1
−4	−1	−1	1	2	7	1
−7	−2	−3	−2	1	6	1
−10	−3	−5	−5	0	5	1
55	22	55	110	55	110	11
·647	·564	·486	·416	·357	·316	·302

$N = 13$

0	1	2	3	4	5	6	7
8	25	44	19	32	13	20	1
7	22	39	17	29	12	19	1
6	19	34	15	26	11	18	1
5	16	29	13	23	10	17	1
4	13	24	11	20	9	16	1
3	10	19	9	17	8	15	1
2	7	14	7	14	7	14	1
1	4	9	5	11	6	13	1
0	1	4	3	8	5	12	1
−1	−2	−1	1	5	4	11	1
−2	−5	−6	−1	2	3	10	1
−3	−8	−11	−3	−1	2	9	1
−4	−11	−16	−5	−4	1	8	1
26	91	182	91	182	91	182	13
·588	·524	·463	·406	·355	·314	·287	·277

$N = 15$

0	1	2	3	4	5	6	7	8
28	29	91	161	35	119	49	77	1
25	26	82	146	32	110	46	74	1
22	23	73	131	29	101	43	71	1
19	20	64	116	26	92	40	68	1
16	17	55	101	23	83	37	65	1
13	14	46	86	20	74	34	62	1
10	11	37	71	17	65	31	59	1
7	8	28	56	14	56	28	56	1
4	5	19	41	11	47	25	53	1
1	2	10	26	8	38	22	50	1
−2	−1	1	11	5	29	19	47	1
−5	−4	−8	−4	2	20	16	44	1
−8	−7	−17	−19	−1	11	13	41	1
−11	−10	−26	−34	−4	2	10	38	1
−14	−13	−35	−49	−7	−7	7	35	1
105	120	420	840	210	840	420	840	15
·543	·492	·442	·395	·352	·314	·285	·265	·258

N = 17

0	1	2	3	4	5	6	7	8	9
32	11	80	12	64	14	16	20	32	1
29	10	73	11	59	13	15	19	31	1
26	9	66	10	54	12	14	18	30	1
23	8	59	9	49	11	13	17	29	1
20	7	52	8	44	10	12	16	28	1
17	6	45	7	39	9	11	15	27	1
14	5	38	6	34	8	10	14	26	1
11	4	31	5	29	7	9	13	25	1
8	3	24	4	24	6	8	12	24	1
5	2	17	3	19	5	7	11	23	1
2	1	10	2	14	4	6	10	22	1
−1	0	3	1	9	3	5	9	21	1
−4	−1	−4	0	4	2	4	8	20	1
−7	−2	−11	−1	−1	1	3	7	19	1
−10	−3	−18	−2	−6	0	2	6	18	1
−13	−4	−25	−3	−11	−1	1	5	17	1
−16	−5	−32	−4	−16	−2	0	4	16	1
136	51	408	68	408	102	136	204	408	17
·507	·464	·423	·383	·347	·313	·284	·262	·248	·243

N = 19

0	1	2	3	4	5	6	7	8	9	10
12	37	51	93	14	15	33	19	24	39	1
11	34	47	86	13	14	31	18	23	38	1
10	31	43	79	12	13	29	17	22	37	1
9	28	39	72	11	12	27	16	21	36	1
8	25	35	65	10	11	25	15	20	35	1
7	22	31	58	9	10	23	14	19	34	1
6	19	27	51	8	9	21	13	18	33	1
5	16	23	44	7	8	19	12	17	32	1
4	13	19	37	6	7	17	11	16	31	1
3	10	15	30	5	6	15	10	15	30	1
2	7	11	23	4	5	13	9	14	29	1
1	4	7	16	3	4	11	8	13	28	1
0	1	3	9	2	3	9	7	12	27	1
−1	−2	−1	2	1	2	7	6	11	26	1
−2	−5	−5	−5	0	1	5	5	10	25	1
−3	−8	−9	−12	−1	0	3	4	9	24	1
−4	−11	−13	−19	−2	−1	1	3	8	23	1
−5	−14	−17	−26	−3	−2	−1	2	7	22	1
−6	−17	−21	−33	−4	−3	−3	1	6	21	1
57	190	285	570	95	114	285	190	285	570	19
·478	·441	·406	·372	·340	·311	·284	·262	·244	·233	·229

$$N = 21$$

0	1	2	3	4	5	6	7	8	9	10	11
40	41	380	175	320	145	52	115	200	85	140	1
37	38	353	163	299	136	49	109	191	82	137	1
34	35	326	151	278	127	46	103	182	79	134	1
31	32	299	139	257	118	43	97	173	76	131	1
28	29	272	127	236	109	40	91	164	73	128	1
25	26	245	115	215	100	37	85	155	70	125	1
22	23	218	103	194	91	34	79	146	67	122	1
19	20	191	91	173	82	31	73	137	64	119	1
16	17	164	79	152	73	28	67	128	61	116	1
13	14	137	67	131	64	25	61	119	58	113	1
10	11	110	55	110	55	22	55	110	55	110	1
7	8	83	43	89	46	19	49	101	52	107	1
4	5	56	31	68	37	16	43	92	49	104	1
1	2	29	19.	47	28	13	37	83	46	101	1
−2	−1	2	7	26	19	10	31	74	43	98	1
−5	−4	−25	−5	5	10	7	25	65	40	95	1
−8	−7	−52	−17	−16	1	4	19	56	37	92	1
−11	−10	−79	−29	−37	−8	1	13	47	34	89	1
−14	−13	−106	−41	−58	−17	−2	7	38	31	86	1
−17	−16	−133	−53	−79	−26	−5	1	29	28	83	1
−20	−19	−160	−65	−100	−35	−8	−5	20	25	80	1
210	231	2310	1155	2310	1155	462	1155	2310	1155	2310	21
·453	·421	·391	·362	·334	·307	·283	·262	·244	·230	·221	·218

QUADRATIC

N = 5

0	1	2	3	2	1	0
9	31	9	−3	−5	3	3
0	9	13	12	6	−5	−3
−4	−3	12	17	12	−3	−4
−3	−5	6	12	13	9	0
3	3	−5	−3	9	31	9
5	35	35	35	35	35	5
2·145	·941	·609	·697	·609	·941	2·145

N = 7

0	1	2	3	4
9	32	5	1	−2
3	15	4	3	3
−1	3	3	4	6
−3	−4	2	4	7
−3	−6	1	3	6
−1	−3	0	1	3
3	5	−1	−2	−2
7	42	14	14	21
1·558	·873	·535	·535	·577

N = 9

0	1	2	3	4	5
42	109	126	378	14	−21
21	63	92	441	273	14
5	27	63	464	447	39
−6	1	39	447	536	54
−12	−15	20	390	540	59
−13	−21	6	293	459	54
−9	−17	−3	156	293	39
0	−3	−7	−21	42	14
14	21	−6	−238	−294	−21
42	165	330	2310	2310	231
1·273	·813	·528	·448	·482	·505

N = 11

0	1	2	3	4	5	6
135	83	270	450	55	−15	−36
81	54	199	414	87	46	9
37	30	138	373	109	92	44
3	11	87	327	121	123	69
−21	−3	46	276	123	139	84
−35	−12	15	220	115	140	89
−39	−16	−6	159	97	126	84
−33	−15	−17	93	69	97	69
−17	−9	−18	22	31	53	44
9	2	−9	−54	−17	−6	9
45	18	10	−135	−75	−80	−36
165	143	715	2145	715	715	429
1·098	·762	·528	·417	·411	·441	·455

$N = 13$

0	1	2	3	4	5	6	7
99	47	33	231	121	33	−33	−11
66	33	25	198	132	77	33	0
38	21	18	167	138	111	86	9
15	11	12	138	139	135	126	16
−3	3	7	111	135	149	153	21
−16	−3	3	86	126	153	167	24
−24	−7	0	63	112	147	168	25
−27	−9	−2	42	93	131	156	24
−25	−9	−3	23	69	105	131	21
−18	−7	−3	6	40	69	93	16
−6	−3	−2	−9	6	23	42	9
11	3	0	−22	−33	−33	−22	0
33	11	3	−33	−77	−99	−99	−11
143	91	91	1001	1001	1001	1001	143
·979	·719	·524	·408	·373	·386	·408	·418

$N = 15$

0	1	2	3	4	5	6	7	8
273	158	819	7 371	2 275	2 184	273	−1 183	−78
195	117	638	6 201	2 184	2 795	1 482	429	−13
127	81	477	5 126	2 073	3 261	2 471	1 776	42
69	50	336	4 146	1 942	3 582	3 240	2 858	87
21	24	215	3 261	1 791	3 758	3 789	3 675	122
−17	3	114	2 471	1 620	3 789	4 118	4 227	147
−45	−13	33	1 776	1 429	3 675	4 227	4 514	162
−63	−24	−28	1 176	1 218	3 416	4 116	4 536	167
−71	−30	−69	671	987	3 012	3 785	4 293	162
−69	−31	−90	261	736	2 463	3 234	3 785	147
−57	−27	−91	−54	465	1 769	2 463	3 012	122
−35	−18	−72	−274	174	930	1 472	1 974	87
−3	−4	−33	−399	−137	−54	261	671	42
39	15	26	−429	−468	−1 183	−1 170	−897	−13
91	39	105	−364	−819	−2 457	−2 821	−2 730	−78
455	340	2 380	30 940	15 470	30 940	30 940	30 940	1 105
·891	·682	·518	·407	·354	·349	·365	·382	·389

N = 17

0	1	2	3	4	5	6	7	8	9
36	409	1 260	308	628	372	52	−20	−156	−21
27	315	1 006	259	573	394	80	111	7	−6
19	231	777	214	519	408	103	222	147	7
12	157	573	173	466	414	121	313	264	18
6	93	394	136	414	412	134	384	358	27
1	39	240	103	363	402	142	435	429	34
−3	−5	111	74	313	384	145	466	477	39
−6	−39	7	49	264	358	143	477	502	42
−8	−63	−72	28	216	324	136	468	504	43
−9	−77	−126	11	169	282	124	439	483	42
−9	−81	−155	−2	123	232	107	390	439	39
−8	−75	−159	−11	78	174	85	321	372	34
−6	−59	−138	−16	34	108	58	232	282	27
−3	−33	−92	−17	−9	34	26	123	169	18
1	3	−21	−14	−51	−48	−11	−6	33	7
6	49	75	−7	−92	−138	−53	−155	−126	−6
12	105	196	4	−132	−236	−100	−324	−308	−21
68	969	3 876	1 292	3 876	3 876	1 292	3 876	3 876	323
·822	·650	·509	·407	·347	·326	·332	·347	·360	·365

N = 19

0	1	2	3	4	5	6	7	8	9	10
459	257	612	7956	1921	765	306	34	−459	−1 377	−136
357	204	499	6 732	1 717	748	357	68	527	−204	−51
265	156	396	5 603	1 523	726	398	97	1 383	824	24
183	113	303	4 569	1 339	699	429	121	2 109	1 707	89
111	75	220	3 630	1 165	667	450	140	2 705	2 445	144
49	42	147	2 786	1 001	630	461	154	3 171	3 038	189
−3	14	84	2 037	847	588	462	163	3 507	3 486	224
−45	−9	31	1 383	703	541	453	167	3 713	3 789	249
−77	−27	−12	824	569	489	434	166	3 789	3 947	264
−99	−40	−45	360	445	432	405	160	3 735	3 960	269
−111	−48	−68	−9	331	370	366	149	3 551	3 828	264
−113	−51	−81	−283	227	303	317	133	3 237	3 551	249
−105	−49	−84	−462	133	231	258	112	2 793	3 127	224
−87	−42	−77	−546	49	154	189	86	2 219	2 562	189
−59	−30	−60	−535	−25	72	110	55	1 515	1 852	144
−21	−13	−33	−429	−89	−15	21	19	681	993	89
27	9	4	−228	−143	−107	−78	−22	−283	−9	24
85	36	51	68	−187	−204	−187	−68	−1 377	−1 156	−51
153	68	108	459	−221	−306	−306	−119	−2 601	−2 448	−136
969	665	1 995	33 915	11 305	6 783	4 845	1 615	33 915	33 915	2 261
·768	·622	·500	·406	·344	·314	·308	·318	·331	·341	·345

$$N = 21$$

0	1	2	3	4	5	6	7	8	9	10	11
285	631	2 565	38 475	29 165	20 805	2 679	6 935	1 425	−3 135	−6 745	−171
228	513	2 129	32 832	25 878	19 589	2 793	9 006	4 712	1 083	1 881	−76
176	405	1 728	27 599	22 776	18 363	2 872	10 767	7 584	4 811	2 448	9
129	307	1 362	22 776	19 859	17 127	2 916	12 218	10 041	8 049	6 242	84
87	219	1 031	18 363	17 127	15 881	2 925	13 359	12 083	10 797	9 501	149
50	141	735	14 360	14 580	14 625	2 899	14 190	13 710	13 055	12 225	204
18	73	474	10 767	12 218	13 359	2 838	14 711	14 922	14 823	14 414	249
−9	15	248	7 584	10 041	12 083	2 742	14 922	15 719	16 101	16 068	284
−31	−33	57	4 811	8 049	10 797	2 611	14 823	16 101	16 889	17 187	309
−48	−71	−99	2 448	6 242	9 501	2 445	14 414	16 068	17 187	17 771	324
−60	−99	−220	495	4 620	8 195	2 244	13 695	15 620	16 995	17 820	329
−67	−117	−306	−1 048	3 183	6 879	2 008	12 666	14 757	16 313	17 334	324
−69	−125	−357	−2 181	1 931	5 553	1 737	11 327	13 479	15 141	16 313	309
−66	−123	−373	−2 904	864	4 217	1 431	9 678	11 786	13 479	14 757	284
−58	−111	−354	−3 217	−18	2 871	1 090	7 719	9 678	11 327	12 666	249
−45	−89	−300	−3 120	−715	1 515	714	5 450	7 155	8 685	10 040	204
−27	−57	−211	−2 613	−1 227	149	303	2 871	4 217	5 553	6 879	149
−4	−15	−87	−1 696	−1 554	−1 227	−143	−18	864	1 931	3 183	84
24	37	72	−369	−1 696	−2 613	−624	−3 217	−2 904	−2 181	−1 048	9
57	99	266	1 368	−1 653	−4 009	−1 140	−6 726	−7 087	−6 783	−5 814	−76
95	171	495	3 515	−1 425	−5 415	−1 691	−10 545	−11 685	−11 875	−11 115	−171
665	1 771	8 855	168 245	168 245	168 245	33 649	168 245	168 245	168 245	168 245	3 059
·722	·597	·490	·405	·344	·307	·294	·296	·306	·317	·325	·328

CUBIC

N = 5

0	1	2	3	2	1	0
16	69	2	-3	2	-1	-4
-14	4	27	12	-8	4	11
-4	-6	12	17	12	-6	-4
11	4	-8	12	27	4	-14
-4	-1	2	-3	2	69	16
5	70	35	35	35	70	5
4·919	·993	·878	·697	·878	·993	4·919

N = 7

0	1	2	3	4
16	39	8	-4	-2
-4	8	19	16	3
-8	-4	16	19	6
-3	-4	6	12	7
4	1	-4	2	6
6	4	-7	-4	3
-4	-2	4	1	-2
7	42	42	42	21
2·903	·964	·673	·673	·577

N = 9

0	1	2	3	4	5
224	85	56	-28	-56	-21
14	28	65	392	84	14
-76	-2	56	515	144	39
-81	-12	36	432	145	54
-36	-9	12	234	108	59
24	0	-9	12	54	54
64	8	-20	-143	4	39
49	8	-14	-140	-21	14
-56	-7	16	112	0	-21
126	99	198	1 386	462	231
2·162	·927	·573	·610	·560	·505

N = 11

0	1	2	3	4	5	6
96	113	48	24	-72	-51	-36
24	48	41	96	132	36	9
-16	8	32	123	232	86	44
-31	-12	22	116	251	106	69
-28	-17	12	86	212	103	84
-14	-12	3	44	138	84	89
4	-2	-4	1	52	56	84
19	8	-8	-32	-23	26	69
24	13	-8	-44	-64	1	44
12	8	-3	-24	-48	-12	9
-24	-12	8	39	48	-6	-36
66	143	143	429	858	429	429
1·771	·889	·535	535	·541	·490	·455

N = 13

0	1	2	3	4	5	6	7
176	265	33	231	− 33	− 407	− 110	− 11
66	132	25	396	132	308	33	0
− 4	42	18	460	222	738	128	9
− 41	− 12	12	444	251	932	182	16
− 52	− 37	7	369	233	939	202	21
− 44	− 40	3	256	182	808	195	24
− 24	− 28	0	126	112	588	168	25
1	− 8	− 2	0	37	328	128	24
24	13	− 3	− 101	− 29	77	82	21
38	28	− 3	− 156	− 72	− 116	37	16
36	30	− 2	− 144	− 78	− 202	0	9
11	12	0	− 44	− 33	− 132	− 22	0
− 44	− 33	3	165	77	143	− 22	− 11
143	364	91	2 002	1 001	4 004	1 001	143
1·526	·853	·524	·479	·501	·484	·441	·418

N = 15

1	2	3	4	5	6	7	8
2 059	8 008	44 044	2 002	− 19 201	− 28 756	− 27 846	− 78
1 144	5 833	52 624	17 017	19 604	8 879	1 404	− 13
484	4 048	54 709	25 762	44 294	34 244	22 599	42
44	2 618	51 524	29 252	57 004	49 054	36 684	87
− 211	1 508	44 294	28 502	59 869	55 024	44 604	122
− 316	683	34 244	24 527	55 024	53 869	47 304	147
− 308	108	22 599	18 342	44 604	47 304	45 729	162
− 216	− 252	10 584	10 962	30 744	37 044	40 824	167
− 81	− 432	− 576	3 402	15 579	24 804	33 534	162
64	− 467	− 9 656	− 3 323	1 244	12 299	24 804	147
184	− 392	− 15 431	− 8 198	− 10 126	1 244	15 579	122
244	− 242	− 16 676	− 10 208	− 16 396	− 6 646	6 804	87
209	− 52	− 12 166	− 8 338	− 15 431	− 9 656	− 576	42
44	143	− 676	− 1 573	− 5 096	− 6 071	− 5 616	− 13
− 286	308	19 019	11 102	16 744	5 824	− 7 371	− 78
3 060	21 420	278 460	139 230	278 460	278 460	278 460	1 105
·821	·522	·443	·458	·464	·440	·405	·389

$N = 17$

0	1	2	3	4	5	6	7	8	9
64	605	1 456	728	208	−132	−320	−384	−352	−21
34	364	1 055	728	468	268	121	20	−42	−6
12	182	728	691	624	534	428	313	196	7
−3	52	468	624	691	684	618	508	369	18
−12	−33	268	534	684	736	708	618	484	27
−16	−80	121	428	618	708	715	656	548	34
−16	−96	20	313	508	618	656	635	568	39
−13	−88	−42	196	369	484	548	568	551	42
−8	−63	−72	84	216	324	408	468	504	43
−2	−28	−77	−16	64	156	253	348	434	42
4	10	−64	−97	−72	−2	100	221	348	39
9	44	−40	−152	−177	−132	−34	100	253	34
12	67	−12	−174	−236	−216	−132	−2	156	27
12	72	13	−156	−234	−236	−177	−72	64	18
8	52	28	−91	−156	−174	−152	−97	−16	7
−1	0	26	28	13	−12	−40	−64	−77	−6
−16	−91	0	208	288	268	176	40	−112	−21
68	969	3 876	3 876	3 876	3 876	3 876	3 876	3 876	323
1·231	·790	·522	·422	·422	·436	·429	·405	·377	·365

$N = 19$

0	1	2	3	4	5	6	7	8	9	10
3 264	14 467	1 632	15 504	25 296	−153	−1 224	−918	−27 744	−561	−136
1 904	9 248	1 193	14 144	36 856	5 372	714	68	−3 264	−136	−51
864	5 168	832	12 601	43 696	9 162	2 104	797	15 424	204	24
109	2 108	542	10 924	46 439	11 427	3 009	1 297	28 901	466	89
−396	−51	316	9 162	45 708	12 377	3 492	1 596	37 748	657	144
−686	−1 428	147	7 364	42 126	12 222	3 616	1 722	42 546	784	189
−796	−2 142	28	5 579	36 316	11 172	3 444	1 703	43 876	854	224
−761	−2 312	−48	3 856	28 901	9 437	3 039	1 567	42 319	874	249
−616	−2 057	−88	2 244	20 504	7 227	2 464	1 342	38 456	851	264
−396	−1 496	−99	792	11 748	4 752	1 782	1 056	32 868	792	269
−136	−748	−88	−451	3 256	2 222	1 056	737	26 136	704	264
129	68	−62	−1 436	−4 349	−153	349	413	18 841	594	249
364	833	−28	−2 114	−10 444	−2 163	−276	112	11 564	469	224
534	1 428	7	−2 436	−14 406	−3 598	−756	−138	4 886	336	189
604	1 734	36	−2 353	−15 612	−4 248	−1 028	−309	−612	202	144
539	1 632	52	−1 816	−13 439	−3 903	−1 029	−373	−4 349	74	89
304	1 003	48	−776	−7 264	−2 353	−696	−302	−5 744	−41	24
−136	−272	17	816	3 536	612	34	−68	−4 216	−136	−51
−816	−2 312	−48	3 009	19 584	5 202	1 224	357	816	−204	−136
3 876	24 871	4 389	74 613	298 452	74 613	21 318	10 659	298 452	6 783	2 261
1·133	·763	·521	·411	·394	·407	·412	·400	·377	·354	·345

$N = 21$

0	1	2	3	4	5	6	7	8	9	10	11
912	5 781	1 938	22 287	32 946	15 447	−9 804	−20 634	−25 764	−52 383	−2 546	−171
570	3 876	1 437	19 380	38 760	46 740	11 571	2 964	−2 850	−12 540	−855	−76
300	2 346	1 020	16 593	41 820	68 610	27 276	20 775	14 844	19 050	540	9
95	1 156	680	13 940	42 469	82 100	37 906	33 380	27 815	43 108	1 660	84
−52	271	410	11 435	41 050	88 253	44 056	41 360	36 560	60 355	2 526	149
−148	−344	203	9 092	37 906	88 112	46 321	45 296	41 576	71 512	3 159	204
−200	−724	52	6 925	33 380	82 720	45 296	45 769	43 360	77 300	3 580	249
−215	−904	−50	4 948	27 815	73 120	41 576	43 360	42 409	78 440	3 810	284
−200	−919	−110	3 175	21 554	60 355	35 756	38 650	39 220	75 653	3 870	309
−162	−804	−135	1 620	14 940	45 468	28 431	32 220	34 290	69 660	3 781	324
−108	−594	−132	297	8 316	29 502	20 196	24 651	28 116	61 182	3 564	329
−45	−324	−108	−780	2 025	13 500	11 646	16 524	21 195	50 940	3 240	324
20	−29	−70	−1 597	−3 590	−1 495	3 376	8 420	14 024	39 655	2 830	309
80	256	−25	−2 140	−8 186	−14 440	−4 019	920	7 100	28 048	2 355	284
128	496	20	−2 395	−11 420	−24 292	−9 944	−5 395	920	16 840	1 836	249
157	656	58	−2 348	−12 949	−30 008	−13 804	−9 944	−4 019	6 752	1 294	204
160	701	82	−1 985	−12 430	−30 545	−15 004	−12 146	−7 220	−1 495	750	149
130	596	85	−1 292	−9 520	−24 860	−12 949	−11 420	−8 186	−7 180	225	84
60	306	60	−255	−3 876	−11 910	−7 044	−7 185	−6 420	−9 582	−260	9
−57	−204	0	1 140	4 845	9 348	3 306	1 140	−1 425	−7 980	−684	−76
−228	−969	−102	2 907	16 986	39 957	18 696	14 136	7 296	−1 653	−1 026	−171
1 197	10 626	5 313	100 947	302 841	605 682	302 841	302 841	302 841	605 682	33 649	3 059
1·055	·738	·520	·405	·374	·382	·391	·389	·374	·353	·335	·328

QUARTIC

N = 7

0	1	2	3	4
25	456	25	−35	5
−25	25	356	155	−30
−5	−35	155	212	75
15	10	−60	150	131
7	20	−65	25	75
−15	−19	70	−65	−30
5	5	−19	20	5
7	462	462	462	231
6·059	·993	·878	·677	·753

N = 9

0	1	2	3	4	5
50	1 231	350	−250	−50	15
−25	175	1 412	1 025	75	−55
−25	−125	1 025	1 112	225	30
0	−75	225	675	304	135
18	45	−330	180	270	179
15	81	−360	−105	135	135
−5	5	37	−110	−35	30
−20	−85	385	37	−120	−55
10	35	−170	10	54	15
18	1 287	2 574	2 574	858	429
3·859	·978	·741	·657	·595	·646

N = 11

0	1	2	3	4	5	6
75	131	30	−30	−45	−15	18
−15	30	59	150	75	0	−45
−35	−10	50	177	125	50	−10
−20	−15	25	125	127	100	60
4	−5	0	50	100	127	120
20	6	−15	−10	60	120	148
20	10	−16	−35	20	80	120
5	5	−5	−23	−10	20	60
−15	−5	10	10	−23	−35	−10
−21	−10	15	30	−15	−48	−45
15	6	−10	−15	15	30	18
33	143	143	429	429	429	429
2·908	·957	·642	·642	·544	·544	·577

N = 13

0	1	2	3	4	5	6	7
275	2 698	825	−825	−3 795	−2 915	−286	110
0	825	1 048	9 900	6 270	2 255	−495	−198
−100	−75	900	12 428	10 140	5 745	640	−135
−95	−345	570	10 140	9 992	7 625	2 230	110
−41	−265	205	5 745	7 625	8 042	3 610	390
20	−52	−90	1 280	4 460	7 220	4 339	600
60	140	−252	−1 890	1 540	5 460	4 200	677
65	220	−260	−3 072	−470	3 140	3 200	600
35	160	−135	−2 245	−1 283	715	1 570	390
−16	−5	60	−60	−990	−1 283	−235	110
−60	−177	220	2 160	−60	−2 245	−1 536	−135
−55	−195	198	2 420	660	−1 485	−1 430	−198
55	165	−195	−1 947	−55	1 760	1 210	110
143	3 094	3 094	34 034	34 034	34 034	17 017	2 431
2·379	·934	·582	·604	·542	·486	·505	·528

$N = 15$

0	1	2	3	4	5	6	7	8
5 005	9 626	25 025	25 025	−50 050	−1 251 250	−749 749	−45 045	270 270
715	3 575	24 482	260 975	89 375	1 011 725	177 320	−420 849	−360 360
−1 265	275	20 075	331 466	147 950	2 240 975	1 038 785	−26 730	−370 062
−1 705	−1 100	13 750	295 900	151 708	2 698 300	1 732 390	746 460	−20 790
−1 249	−1 250	7 075	203 725	122 650	2 614 126	2 187 505	1 586 115	473 130
−415	−749	1 240	94 435	78 745	2 187 505	2 365 126	2 257 875	945 000
405	−45	−2 943	−2 430	33 930	1 586 115	2 257 875	2 605 626	1 275 750
945	540	−5 040	−67 284	−1 890	946 260	1 890 000	2 551 500	1 393 938
1 065	810	−4 995	−90 495	−22 842	372 870	1 317 375	2 095 875	1 275 750
751	695	−3 130	−72 385	−27 085	−60 499	627 500	1 317 375	945 000
115	251	−145	−23 230	−16 810	−311 665	−60 499	372 870	473 130
−605	−340	2 882	36 740	1 760	−369 820	−595 870	−502 524	−20 790
−1 045	−770	4 495	77 341	18 370	−255 530	−796 235	−995 445	−370 062
−715	−605	2 860	58 435	18 733	−20 735	−447 590	−714 285	−360 360
1 001	715	−4 235	−70 070	−15 470	251 251	695 695	810 810	270 270
3 003	11 628	81 396	1 058 148	529 074	11 639 628	11 639 628	11 639 628	5 819 814
2·040	·910	·548	·560	·535	·474	·451	·473	·489

$N = 17$

0	1	2	3	4	5	6	7	8	9
1 300	761	1 300	260	−260	−2 730	−4 628	−2 340	260	39(
325	325	1 094	845	585	2 210	1 690	−403	−1 755	−39(
−195	65	845	1 042	975	4 875	5 915	2 080	−1 079	−52(
−390	−65	585	975	1 042	5 850	8 385	4 615	1 170	−23⋅
−372	−105	340	750	900	5 648	9 420	6 810	4 060	27(
−235	−89	130	455	645	4 710	9 322	8 375	6 845	83(
−55	−45	−31	160	355	3 405	8 375	9 122	8 965	1 32(
110	5	−135	−83	90	2 030	6 845	8 965	10 046	1 65(
220	45	−180	−240	−108	810	4 980	7 920	9 900	1 76(
253	65	−170	−295	−215	−102	3 010	6 105	8 525	1 65(
205	61	−115	−250	−225	−625	1 147	3 740	6 105	1 32(
90	35	−31	−125	−150	−750	−415	1 147	3 010	83(
−60	−5	60	42	−20	−540	−1 500	−1 250	−204	27(
−195	−45	130	195	117	−130	−1 950	−2 925	−2 795	−23⋅
−247	−65	145	260	195	273	−1 625	−3 250	−3 835	−52(
−130	−39	65	145	130	390	−403	−1 495	−2 210	−39(
260	65	−156	−260	−180	−130	1 820	3 172	3 380	39(
884	969	3 876	3 876	3 876	25 194	50 388	50 388	50 388	8 39?
1·804	·886	·531	·518	·518	·473	·430	·425	·447	·459

$$N = 19$$

0	1	2	3	4	5	6	7	8	9	10
2 550	3 583	5 100	25 500	−17 850	−47 430	−50 184	−36 720	−15 810	255	340
850	1 700	3 949	49 300	67 150	40 460	18 870	2 516	−8 670	−680	−255
−150	500	2 900	57 533	109 450	89 640	62 360	33 170	6 466	−660	−420
−625	−175	1 975	54 725	120 266	109 455	85 515	55 585	25 445	−22	−290
−729	−465	1 190	44 820	109 455	108 188	93 060	70 230	44 795	945	18
−595	−492	555	31 180	85 515	93 060	89 216	77 700	61 725	2 000	405
−335	−360	74	16 585	55 585	70 230	77 700	78 716	74 125	2 950	790
−40	−155	−255	3 233	25 445	44 795	61 725	74 125	80 566	3 650	1 110
220	55	−440	−7 260	−484	20 790	44 000	64 900	80 300	4 003	1 320
396	220	−495	−13 860	−19 140	1 188	26 730	52 140	73 260	3 960	1 393
460	308	−440	−16 115	−28 820	−12 100	11 616	37 070	60 060	3 520	1 320
405	305	−301	−14 155	−29 180	−18 225	−145	21 041	41 995	2 730	1 110
245	215	−110	−8 692	−21 235	−17 400	−7 860	5 530	21 041	1 685	790
15	60	95	−1 020	−7 359	−10 900	−11 340	−7 860	−145	528	405
−229	−120	270	6 985	8 715	−1 062	−10 900	−17 400	−18 225	−550	18
−410	−267	365	12 865	21 895	8 715	−7 359	−21 235	−29 180	−1 310	−290
−430	−304	324	13 580	25 730	13 970	−2 040	−17 384	−28 310	−1 465	−420
−170	−140	85	5 508	12 410	9 180	3 230	−3 740	−10 234	−680	−255
510	340	−420	−15 555	−27 234	−12 240	6 120	21 930	31 110	1 428	340
1 938	4 807	14 421	245 157	490 314	490 314	490 314	490 314	490 314	22 287	7 429
1·630	·863	·523	·484	·495	·470	·427	·401	·405	·424	·433

$N = 21$

0	1	2	3	4	5	6	7	8	9	10	11
24 225	18 813	1938	13 566	−1 938	−405 042	−2 603 703	−464 151	−307 173	−109 497	9 044	127 908
9 690	9 690	1437	19 380	38 760	397 290	983 535	50 388	−48 450	−106 590	−14 535	−71 060
510	3 570	1020	21 183	60 180	865 470	3 250 230	412 845	183 498	−14 790	−18 360	−143 055
−4 505	−170	680	20 060	66 949	1 074 230	4 464 370	647 020	381 055	130 798	−8 500	−123 420
−6 419	−2 090	410	16 970	63 190	1 090 553	4 868 365	775 075	538 495	299 920	9 732	−43 340
−6 170	−2 687	203	12 746	52 522	973 673	4 679 047	817 534	651 982	467 173	31 779	70 158
−4 570	−2 395	52	8 095	38 060	775 075	4 087 670	793 283	719 570	612 005	53 840	194 205
−2 305	−1 585	−50	3 598	22 415	538 495	3 259 910	719 570	741 203	718 715	72 870	310 090
65	−565	−110	−290	7 694	299 920	2 335 865	612 005	718 715	776 453	86 580	403 260
2 106	420	−135	−3 240	−4 500	87 588	1 430 055	484 560	655 830	779 220	93 437	463 320
3 510	1 188	−132	−5 049	−13 068	−78 012	631 422	349 569	558 162	725 868	92 664	484 033
4 095	1 620	−108	−5 640	−17 415	−184 140	3 330	217 728	433 215	620 100	84 240	463 320
3 805	1 660	−70	−5 062	−17 450	−225 805	−416 435	98 095	290 383	470 470	68 900	403 260
2 710	1 315	−25	−3 490	−13 586	−205 765	−615 665	−1 910	140 950	290 383	48 135	310 090
1 006	655	20	−1 225	−6 740	−134 527	−607 730	−76 505	−1 910	98 095	24 192	194 205
−985	−187	58	1 306	1 667	−30 347	−431 578	−121 546	−123 133	−83 287	74	70 158
−2 815	−1 015	82	3 550	9 710	80 770	−151 735	−134 527	−205 765	−225 805	−20 460	−43 340
−3 910	−1 570	85	4 828	14 960	165 070	141 695	−114 580	−230 962	−296 650	−32 895	−123 420
−3 570	−1 530	60	4 335	14 484	181 050	333 030	−62 475	−177 990	−258 162	−31 960	−143 055
−969	−510	0	1 140	4 845	79 458	281 010	19 380	−24 225	−67 830	−11 628	−71 060
4 845	1 938	−102	−5 814	−17 898	−196 707	−181 203	126 939	254 847	321 708	34 884	127 908
20 349	26 565	5 313	100 947	302 841	5 148 297	25 741 485	5 148 297	5 148 297	5 148 297	572 033	2 860 165
1·494	·842	·520	·458	·470	·460	·426	·393	·379	·388	·404	·411

QUINTIC

N = 7

0	1	2	3	4
36	923	1	−5	5
−69	6	148	30	−30
50	−15	15	233	75
15	20	−20	100	131
−48	−15	15	−75	75
29	6	−6	30	−30
−6	−1	1	−5	5
7	924	154	308	231
15·625	·999	·980	·980	·753

N = 9

0	1	2	3	4	5
48	425	72	−54	32	15
−57	18	1 385	261	−213	−55
−2	−27	522	400	582	30
33	8	−213	291	905	135
12	15	−220	60	540	179
−23	−6	123	−101	−27	135
−18	−13	186	−66	−202	30
27	12	−187	93	123	−55
−8	−3	48	−26	−24	15
12	429	1 716	858	1 716	429
7·654	·995	·898	·683	·726	·646

N = 11

0	1	2	3	4	5	6
108	557	27	−153	−4	6	18
−81	54	184	666	3	−22	−45
−46	−51	111	719	38	13	−10
24	−16	6	456	57	48	60
48	24	−44	156	48	57	120
20	24	−30	−40	20	40	143
−24	−4	12	−96	−8	12	120
−39	−24	34	−48	−18	−8	60
−4	−9	9	29	−4	−8	−10
45	26	−36	54	17	6	−45
−18	−9	13	−27	−6	−1	18
33	572	286	1 716	143	143	429
5·078	·987	·802	·647	·631	·631	·577

N = 13

0	1	2	3	4	5	6	7
792	420	132	−429	−704	33	132	110
−363	66	455	1 881	1 221	−352	−330	−198
−398	−39	342	2 030	2 586	453	−50	−135
−69	−32	111	1 293	3 045	1 314	405	110
204	3	−64	453	2 628	1 680	720	390
260	24	−120	−100	1 620	1 440	777	600
120	20	−72	−270	440	780	600	677
−90	0	20	−156	−480	40	300	600
−216	−18	84	47	−816	−429	20	390
−153	−18	69	147	−473	−408	−120	110
78	3	−22	54	294	47	−78	−135
253	24	−99	−121	759	462	55	−198
−132	−11	48	33	−396	−198	0	110
286	442	884	4 862	9 724	4 862	2 431	2 431
3·849	·975	·717	·646	·560	·588	·565	·528

$$N = 15$$

0	1	2	3	4	5	6	7	8
1 716	11 919	1 287	− 11 583	− 4 147	− 67 353	5 577	22 165	2 145
− 429	2 574	2 742	59 202	7 293	10 582	− 17 160	− 40 326	− 2 860
− 814	− 891	2 277	67 827	11 913	227 337	− 605	− 23 925	− 2 937
− 429	− 1 276	1 122	47 652	12 048	422 532	26 730	22 660	− 165
96	− 471	37	20 667	9 603	513 417	47 835	68 055	3 755
425	390	− 600	− 550	6 075	478 350	55 002	95 250	7 500
450	775	− 705	− 10 875	2 575	340 275	47 625	98 629	10 125
225	600	− 400	− 10 680	− 150	150 200	30 000	81 000	11 063
− 100	75	75	− 3 875	− 1 695	− 29 325	9 125	50 625	10 125
− 345	− 450	450	4 050	− 1 975	− 140 730	− 7 500	18 250	7 500
− 366	− 639	513	7 983	− 1 203	− 147 947	− 14 073	− 5 865	3 755
− 121	− 324	198	5 148	132	− 52 932	− 8 690	− 14 916	− 165
264	341	− 327	− 2 937	1 287	87 813	4 455	− 8 525	− 2 937
429	726	− 572	− 8 502	1 287	146 718	12 870	4 290	− 2 860
− 286	− 429	363	4 433	− 1 053	− 91 377	− 6 435	2 145	2 145
715	12 920	6 460	167 960	41 990	1 847 560	184 756	369 512	46 189
3·137	·960	·652	·635	·536	·527	·546	·517	·489

$$N = 17$$

0	1	2	3	4	5	6	7	8	9
3 744	6 920	1 872	− 52	− 832	− 1 794	− 208	2 236	6 240	195
− 351	1 872	2 825	403	1 443	1 391	− 1 391	− 4 407	− 8 515	− 195
− 1 534	− 312	2 418	486	2 262	3 939	1 066	− 2 496	− 7 878	− 260
− 1 209	− 832	1 443	377	2 201	5 499	4 511	2 899	195	− 117
− 348	− 552	428	202	1 692	5 972	7 276	8 394	10 100	135
443	− 48	− 321	41	1 041	5 457	8 511	12 027	18 255	415
858	344	− 678	− 64	446	4 197	8 018	12 994	22 770	660
825	480	− 655	− 101	15	2 525	6 085	11 385	23 117	825
440	360	− 360	− 80	− 216	810	3 320	7 920	19 800	883
− 99	80	45	− 25	− 265	− 597	485	3 685	14 025	825
− 558	− 216	386	34	− 186	− 1 417	− 1 670	− 132	7 370	660
− 733	− 384	519	69	− 51	− 1 497	− 2 573	− 2 505	1 455	415
− 516	− 328	372	62	68	− 864	− 1 996	− 2 834	− 2 388	135
39	− 48	− 13	13	117	221	− 221	− 1 209	− 3 445	− 117
650	312	− 438	− 52	78	1 209	1 794	1 326	− 1 950	− 260
741	416	− 507	− 73	− 13	1 209	2 249	2 509	585	− 195
− 624	− 312	416	52	− 48	− 1 066	− 1 664	− 1 404	1 040	195
1 768	7 752	7 752	1 292	7 752	25 194	33 592	50 388	100 776	4 199
2·675	·945	·604	·613	·533	·487	·503	·508	·479	·459

$N = 19$

0	1	2	3	4	5	6	7	8	9	10
3 672	8 288	3 978	−3 978	−16 796	−91 494	−19 890	5 066	25 704	1 377	340
102	2 652	4 697	135 252	29 614	78 676	−1 326	−18 020	−36 346	−1 428	−255
−1 228	−78	3 978	167 888	46 904	176 046	33 254	−6 274	−33 420	−1 738	−420
−1 263	−988	2 613	140 712	46 256	216 996	68 289	20 256	1 839	−660	−290
−696	−897	1 157	88 023	36 166	216 475	93 951	47 689	46 016	978	18
−1	−390	−39	33 254	22 763	187 902	105 254	67 442	83 703	2 594	405
534	149	−795	−9 411	10 128	143 067	101 163	75 361	106 278	3 819	790
774	504	−1 069	−33 420	613	92 032	83 703	70 852	110 684	4 464	1 110
704	594	−924	−38 236	−4 840	43 032	57 068	56 012	98 208	4 487	1 320
396	440	−495	−27 720	−6 380	2 376	26 730	34 760	73 260	3 960	1 393
−24	132	44	−8 514	−4 928	−25 652	−1 452	11 968	42 152	3 036	1 320
−409	−204	513	11 576	−1 858	−38 892	−22 123	−7 408	11 877	1 916	1 110
−624	−439	759	25 197	1 322	−37 407	−31 323	−19 197	−11 112	816	790
−579	−474	689	27 066	3 289	−23 582	−27 378	−20 882	−22 123	−66	405
−262	−273	303	15 587	3 224	−2 223	−11 791	−12 469	−19 446	−583	18
228	104	−273	−5 532	1 131	19 344	9 867	2 644	−5 574	−672	−290
648	468	−754	−25 662	−1 844	31 174	27 066	16 798	11 576	−387	−420
578	468	−663	−25 636	−3 094	20 604	23 426	17 204	17 442	68	−255
−612	−442	702	23 868	1 768	−27 846	−24 174	−14 926	−10 404	306	340
1 938	9 614	14 421	490 314	163 438	980 628	490 314	326 876	490 314	22 287	7 429
2·351	·928	·571	·585	·532	·470	·463	·480	·475	·449	·433

$N = 21$

0	1	2	3	4	5	6	7	8	9	10	11
15 504	95 735	69 768	52 326	− 392 768	− 3 858 558	− 10 007 832	− 1 128 562	4 480 656	161 823	173 128	11 628
1 938	34 884	69 376	540 702	742 254	3 335 944	3 164 754	− 2 060 094	− 8 076 938	− 170 544	− 136 629	− 6 460
− 4 012	2 754	56 916	675 137	1 195 644	7 085 634	14 273 404	974 406	− 6 647 748	− 207 434	− 206 244	− 13 005
− 5 202	− 10 336	39 066	597 822	1 213 424	8 388 684	22 392 774	5 579 706	1 988 082	− 77 724	− 128 894	− 11 220
− 3 816	− 11 946	20 656	416 802	986 904	8 065 949	27 104 544	10 061 946	12 841 512	121 356	22 656	− 3 940
− 1 443	− 7 746	4 899	209 903	658 611	6 776 136	28 415 567	13 338 009	22 497 543	320 604	193 974	6 378
846	− 1 747	− 6 378	28 659	328 218	5 030 973	26 676 018	14 844 892	28 898 154	475 227	346 882	17 655
2 421	3 468	− 12 503	− 97 761	58 473	3 210 378	22 497 543	14 449 077	31 125 239	561 222	457 377	28 190
3 016	6 513	− 13 728	− 158 626	− 118 872	1 577 628	16 671 408	12 355 902	29 183 544	571 757	513 552	36 660
2 652	6 968	− 10 998	− 157 716	− 197 132	294 528	10 086 648	9 018 932	23 783 604	513 552	513 517	42 120
1 560	5 148	− 5 720	− 109 395	− 188 760	− 563 420	3 648 216	5 049 330	16 124 680	403 260	463 320	44 003
104	1 872	468	− 34 684	− 119 418	− 991 848	− 1 804 868	1 125 228	7 677 696	263 848	374 868	42 120
− 1 296	− 1 768	5 928	42 666	− 22 048	− 1 042 353	− 5 581 368	− 2 098 902	− 31 824	120 978	263 848	36 660
− 2 241	− 4 638	9 253	100 101	69 057	− 807 328	− 7 219 683	− 4 082 877	− 5 640 819	− 612	147 648	28 190
− 2 430	− 5 793	9 498	120 191	124 182	− 404 793	− 6 569 698	− 4 491 432	− 8 165 754	− 80 727	43 278	17 655
− 1 737	− 4 708	6 411	94 557	124 119	36 774	− 3 874 635	− 3 284 849	− 7 219 683	− 107 334	− 34 709	6 378
− 288	− 1 509	664	27 798	66 096	393 606	147 096	− 809 586	− 3 229 312	− 80 181	− 76 296	− 3 940
1 462	2 796	− 5 916	− 58 582	− 30 294	561 816	4 220 046	2 111 094	2 347 938	− 14 416	− 78 081	− 11 220
2 652	6 086	− 10 116	− 122 247	− 117 164	472 566	6 429 876	4 086 494	6 806 868	55 794	− 45 356	− 13 005
1 938	4 896	− 7 106	− 96 102	− 112 404	107 236	4 141 506	3 067 854	5 977 438	73 644	5 814	− 6 460
− 2 584	− 5 814	9 792	115 634	106 248	− 487 407	− 6 082 736	− 3 742 278	− 5 992 296	− 43 928	46 512	− 11 628
9 044	115 115	230 230	2 187 185	4 374 370	37 182 145	148 728 580	74 364 290	148 728 580	2 860 165	2 860 165	260 015
2·111	·912	·549	·556	·527	·466	·437	·447	·457	·447	·424	·411

APPENDIX B

Table B.1 *Significance points of the Durbin–Watson statistics d_L and d_U: 1 per cent.*

n	$k'=1$ d_L	d_U	$k'=2$ d_L	d_U	$k'=3$ d_L	d_U	$k'=4$ d_L	d_U	$k'=5$ d_L	d_U
15	0·81	1·07	0·70	1·25	0·59	1·46	0·49	1·70	0·39	1·96
16	0·84	1·09	0·74	1·25	0·63	1·44	0·53	1·66	0·44	1·90
17	0·87	1·10	0·77	1·25	0·67	1·43	0·57	1·63	0·48	1·85
18	0·90	1·12	0·80	1·26	0·71	1·42	0·61	1·60	0·52	1·80
19	0·93	1·13	0·83	1·26	0·74	1·41	0·65	1·58	0·56	1·77
20	0·95	1·15	0·86	1·27	0·77	1·41	0·68	1·57	0·60	1·74
21	0·97	1·16	0·89	1·27	0·80	1·41	0·72	1·55	0·63	1·71
22	1·00	1·17	0·91	1·28	0·83	1·40	0·75	1·54	0·66	1·69
23	1·02	1·19	0·94	1·29	0·86	1·40	0·77	1·53	0·70	1·67
24	1·04	1·20	0·96	1·30	0·88	1·41	0·80	1·53	0·72	1·66
25	1·05	1·21	0·98	1·30	0·90	1·41	0·83	1·52	0·75	1·65
26	1·07	1·22	1·00	1·31	0·93	1·41	0·85	1·52	0·78	1·64
27	1·09	1·23	1·02	1·32	0·95	1·41	0·88	1·51	0·81	1·63
28	1·10	1·24	1·04	1·32	0·97	1·41	0·90	1·51	0·83	1·62
29	1·12	1·25	1·05	1·33	0·99	1·42	0·92	1·51	0·85	1·61
30	1·13	1·26	1·07	1·34	1·01	1·42	0·94	1·51	0·88	1·61
31	1·15	1·27	1·08	1·34	1·02	1·42	0·96	1·51	0·90	1·60
32	1·16	1·28	1·10	1·35	1·04	1·43	0·98	1·51	0·92	1·60
33	1·17	1·29	1·11	1·36	1·05	1·43	1·00	1·51	0·94	1·59
34	1·18	1·30	1·13	1·36	1·07	1·43	1·01	1·51	0·95	1·59
35	1·19	1·31	1·14	1·37	1·08	1·44	1·03	1·51	0·97	1·59
36	1·21	1·32	1·15	1·38	1·10	1·44	1·04	1·51	0·99	1·59
37	1·22	1·32	1·16	1·38	1·11	1·45	1·06	1·51	1·00	1·59
38	1·23	1·33	1·18	1·39	1·12	1·45	1·07	1·52	1·02	1·58
39	1·24	1·34	1·19	1·39	1·14	1·45	1·09	1·52	1·03	1·58
40	1·25	1·34	1·20	1·40	1·15	1·46	1·10	1·52	1·05	1·58
45	1·29	1·38	1·24	1·42	1·20	1·48	1·16	1·53	1·11	1·58
50	1·32	1·40	1·28	1·45	1·24	1·49	1·20	1·54	1·16	1·59
55	1·36	1·43	1·32	1·47	1·28	1·51	1·25	1·55	1·21	1·59
60	1·38	1·45	1·35	1·48	1·32	1·52	1·28	1·56	1·25	1·60
65	1·41	1·47	1·38	1·50	1·35	1·53	1·31	1·57	1·28	1·61
70	1·43	1·49	1·40	1·52	1·37	1·55	1·34	1·58	1·31	1·61
75	1·45	1·50	1·42	1·53	1·39	1·56	1·37	1·59	1·34	1·62
80	1·47	1·52	1·44	1·54	1·42	1·57	1·39	1·60	1·36	1·62
85	1·48	1·53	1·46	1·55	1·43	1·58	1·41	1·60	1·39	1·63
90	1·50	1·54	1·47	1·56	1·45	1·59	1·43	1·61	1·41	1·64
95	1·51	1·55	1·49	1·57	1·47	1·60	1·45	1·62	1·42	1·64
100	1·52	1·56	1·50	1·58	1·48	1·60	1·46	1·63	1·44	1·65

Table B.2 *Significance points of the Durbin–Watson statistics d_L and d_U: 5 per cent.*

n	$k' = 1$		$k' = 2$		$k' = 3$		$k' = 4$		$k' = 5$	
	d_L	d_U	d_L	d_U	d_L	d_U	d_L	d_U	d_L	d_U
15	1·08	1·36	0·95	1·54	0·82	1·75	0·69	1·97	0·56	2·21
16	1·10	1·37	0·98	1·54	0·86	1·73	0·74	1·93	0·62	2·15
17	1·13	1·38	1·02	1·54	0·90	1·71	0·78	1·90	0·67	2·10
18	1·16	1·39	1·05	1·53	0·93	1·69	0·82	1·87	0·71	2·06
19	1·18	1·40	1·08	1·53	0·97	1·68	0·86	1·85	0·75	2·02
20	1·20	1·41	1·10	1·54	1·00	1·68	0·90	1·83	0·79	1·99
21	1·22	1·42	1·13	1·54	1·03	1·67	0·93	1·81	0·83	1·96
22	1·24	1·43	1·15	1·54	1·05	1·66	0·96	1·80	0·86	1·94
23	1·26	1·44	1·17	1·54	1·08	1·66	0·99	1·79	0·90	1·92
24	1·27	1·45	1·19	1·55	1·10	1·66	1·01	1·78	0·93	1·90
25	1·29	1·45	1·21	1·55	1·12	1·66	1·04	1·77	0·95	1·89
26	1·30	1·46	1·22	1·55	1·14	1·65	1·06	1·76	0·98	1·88
27	1·32	1·47	1·24	1·56	1·16	1·65	1·08	1·76	1·01	1·86
28	1·33	1·48	1·26	1·56	1·18	1·65	1·10	1·75	1·03	1·85
29	1·34	1·48	1·27	1·56	1·20	1·65	1·12	1·74	1·05	1·84
30	1·35	1·49	1·28	1·57	1·21	1·65	1·14	1·74	1·07	1·83
31	1·36	1·50	1·30	1·57	1·23	1·65	1·16	1·74	1·09	1·83
32	1·37	1·50	1·31	1·57	1·24	1·65	1·18	1·73	1·11	1·82
33	1·38	1·51	1·32	1·58	1·26	1·65	1·19	1·73	1·13	1·81
34	1·39	1·51	1·33	1·58	1·27	1·65	1·21	1·73	1·15	1·81
35	1·40	1·52	1·34	1·58	1·28	1·65	1·22	1·73	1·16	1·80
36	1·41	1·52	1·35	1·59	1·29	1·65	1·24	1·73	1·18	1·80
37	1·42	1·53	1·36	1·59	1·31	1·66	1·25	1·72	1·19	1·80
38	1·43	1·54	1·37	1·59	1·32	1·66	1·26	1·72	1·21	1·79
39	1·43	1·54	1·38	1·60	1·33	1·66	1·27	1·72	1·22	1·79
40	1·44	1·54	1·39	1·60	1·34	1·66	1·29	1·72	1·23	1·79
45	1·48	1·57	1·43	1·62	1·38	1·67	1·34	1·72	1·29	1·78
50	1·50	1·59	1·46	1·63	1·42	1·67	1·38	1·72	1·34	1·77
55	1·53	1·60	1·49	1·64	1·45	1·68	1·41	1·72	1·38	1·77
60	1·55	1·62	1·51	1·65	1·48	1·69	1·44	1·73	1·41	1·77
65	1·57	1·63	1·54	1·66	1·50	1·70	1·47	1·73	1·44	1·77
70	1·58	1·64	1·55	1·67	1·52	1·70	1·49	1·74	1·46	1·77
75	1·60	1·65	1·57	1·68	1·54	1·71	1·51	1·74	1·49	1·77
80	1·61	1·66	1·59	1·69	1·56	1·72	1·53	1·74	1·51	1·77
85	1·62	1·67	1·60	1·70	1·57	1·72	1·55	1·75	1·52	1·77
90	1·63	1·68	1·61	1·70	1·59	1·73	1·57	1·75	1·54	1·78
95	1·64	1·69	1·62	1·71	1·60	1·73	1·58	1·75	1·56	1·78
100	1·65	1·69	1·63	1·72	1·61	1·74	1·59	1·76	1·57	1·78

INDEX

195

DATE DUE

NOV 29 '90			
MAY 8 '92			
JUN 22 '92			
JUN 22 '92			
012097			
	261-2500		Printed in USA